CORE BUSINESS PROGRAM

COMPUTER APPLICATIONS

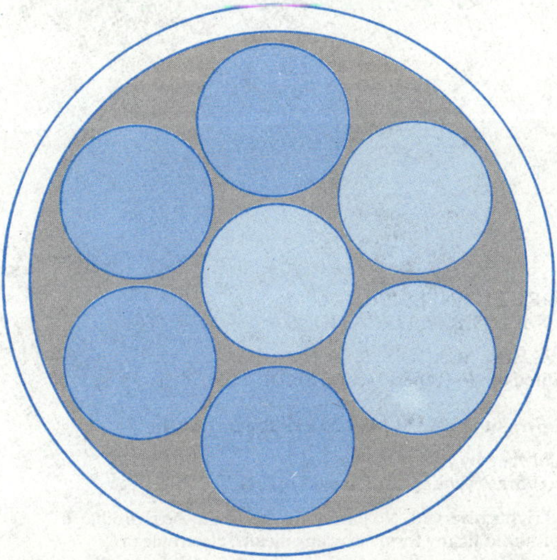

By
J.R. Firth

Edited by
F. Robert Jacobs

Facts On File Publications
460 Park Avenue South
New York, N.Y. 10016

CORE BUSINESS PROGRAM
COMPUTER APPLICATIONS

J.R. Firth, M.A.
Edited by: F. Robert Jacobs, Ph.D.

Published by Facts On File, Inc.
460 Park Avenue South, New York, N.Y. 10016

First edition published in the United Kingdom in 1983 by Mitchell Beazley London Ltd., Mill House,
87-89 Shaftesbury Avenue, London W1V 7AD, England.

Library of Congress Cataloging in Publication Data
Main entry under title:

Computer Applications.
(Core business program)
Includes index.
1. Business–Data processing. 2. Basic (Computer program language) I. Jacobs, F. Robert. II. Title. III. Series.
HF5548.2.C58136 1983 001.64'25 83-5610
ISBN 0-87196-801-0

Printed and Bound in Great Britain
10 9 8 7 6 5 4 3 2 1

Contents

Program list

Introduction

This book is designed to introduce the reader to the BASIC programming language. Throughout the book actual business applications are used to demonstrate the various facets of BASIC. The reader can easily take the examples used in the book and apply them to many problems that are encountered in a business setting. Instead of specializing in one version of this language, which would only be applicable to a particular computer system, the elements of minimal or standard BASIC are considered. Minimal BASIC is that part of the language which is common to all of its different versions and is, in itself, a complete language. Nearly everything that can be programmed in a sophisticated version of BASIC can also be programmed in minimal BASIC.

BASIC is one of the easiest languages to learn but the ability to translate a problem into a set of instructions which the computer can understand depends largely upon the techniques which the programmer can draw upon. For this reason, a large part of the book is concerned with procedures such as sorting and searching, which are fundamental to programming in any computer language. In each chapter a set of new BASIC instructions and techniques is introduced and examples given of their applications in such areas as statistics, operations research, finance, stock control, profit maximization and simulation, etc. The programs become progressively more difficult and each chapter depends upon the concepts introduced in earlier chapters. Mathematical methods are kept to a minimum, wherever possible, and all applications to business studies are explained in the text.

The first chapter gives a brief description of some of the key terms used in computer science. The introduction to the language is given in Chapters 2–9 and 11–12. Chapter 10 surveys the whole process of programming; methods of checking and debugging programs are also given. Chapters 13 and 14 contain much longer programs to solve certain problems in linear programming, queuing, critical path analysis and assignment. In Chapter 7 the mathematical functions available in BASIC are included for the sake of completeness. It is not necessary to understand them in order to program successfully.

Chapter 1

Programming

This chapter gives a brief description of the key terms used in computer science.

COMPUTERS

Computers are machines which process **data** or **information** to produce intelligible results. They can be classified under the following three headings:

Mainframes are large powerful computers. They have a huge storage potential and can process information at a very high rate. The IBM 370 series is one example of a mainframe computer.

Minicomputers are a smaller version of mainframe computers and tend to be used for particular purposes. The DEC PDP 11 series, for example, is often used in scientific laboratories.

Microcomputers are even smaller computers which use specially designed integrated circuits. They are now used extensively in business, schools and in the home because of their low cost. However, they do not have the speed or the storage potential of the mainframes. Examples of these are the IBM–PC and the Apple II.

HARDWARE

Hardware is the term for all the electronic and mechanical components of a computer.

Input devices Information is fed into a computer by typing in data at a keyboard or teletype or by using punched cards and a card reader.

Storage Information can be stored either in the computer's memory or in one of its storage devices using floppy discs, magnetic tapes, etc.

Output devices The results are printed on a visual display unit (VDU), high-speed line printer or a teletype.

Central processor This is the part of the computer which sends signals to the input and output devices and performs all the arithmetic and logic operations. The input, output and storage devices are called **peripheral devices**, i.e. they are external to the main computer containing the central processor and memory.

SOFTWARE

Software is the term for all the programs which may be used to give instructions to the computer.

Systems programs These programs give the user the ability to communicate with the computer and its peripheral devices.

Applications programs These perform specific tasks such as finding the standard deviation of a set of numbers or calculating the internal rate of return of a cash flow project. This book is concerned with these types of programs.

PROGRAMMING LANGUAGES

Applications programs can be written in any of the following languages:

Machine languages Instructions to the computer are written in binary code. Such programs are very long and difficult to follow, because they consist of lines of zeros and ones, but the computer processes them very quickly.

Low level languages These use mnemonics instead of binary numbers. For example, the code 001 in machine language might mean add and would be replaced by the word ADD in a low level language. Low level languages which correspond closely to the machine language are called **assembly languages**. A systems program called an **assembler** automatically translates the assembly language into a machine language so that the computer can understand it.

High level languages These represent a further sophistication in computer language development. It is very time consuming to write a program in a machine or assembly language, so languages were developed which would replace a large set of commands by a single statement. Examples of high level languages are ALGOL, BASIC, COBOL, FORTRAN and PASCAL. A systems program called a **compiler** or an **interpreter** translates the high level language into a machine language.

BASIC (Beginners Allpurpose Symbolic Instruction Code). This language was originally invented for people who were not trained as programmers, but it is now widely used in many applications.

In this book it is assumed the user has access to a computer that is running a BASIC interpreter. The user should consult the instruction manual provided with the computer to find out how to invoke this system program.

Finally, the user should consult his manual to find out how to save BASIC programs on an external storage device. Similarly the user should learn how to load programs. Usually these are simple commands such as SAVE "program name" and LOAD "program name".

Chapter 2

Fundamentals of BASIC

A SIMPLE PROGRAM

As an introduction to the BASIC programming language, consider the following program which calculates simple interest.

$$I = \frac{PR}{100}$$

```
10  PRINT "A PROGRAM TO CALCULATE SIMPLE INTEREST"
20  PRINT "INPUT THE PRINCIPAL IN DOLLARS"
30  INPUT P
40  PRINT "INPUT THE RATE OF INTEREST"
50  INPUT R
60  I = P*R/100
70  PRINT I
80  END
```

The program consists of eight statements, each beginning with a line number which must be a whole number in the range 1 to 99999. The computer executes these statements in numerical order. It is common practice to use line numbers 1Ø, 2Ø, 3Ø, 4Ø, etc., rather than 1, 2, 3, 4, etc., so that any additions to the program may be inserted easily. For example, a new statement may be placed between lines 8Ø and 9Ø by giving it the line number 85. Note that the symbol Ø is used to represent the number zero in order to avoid confusion with the letter O.

Before explaining the terms used in this program, here is a **computer run** showing the output when P = 200, and 1=7.

```
A PROGRAM TO CALCULATE SIMPLE INTEREST
INPUT THE PRINCIPAL IN DOLLARS
? 200
INPUT THE RATE OF INTEREST
? 7
 14
```

The **PRINT** statement on line 1Ø of the program is an instruction to the computer to print on the screen everything after the word PRINT that is enclosed by the quotation marks. This has appeared as the first line of the computer run and is simply a heading which explains the purpose of the program. The computer will not detect any spelling errors inside the quotation marks although, if the word PRINT itself was typed incorrectly, as PRIMT for example, an error message would be indicated by the computer.

Line 2Ø of the program is a similar PRINT command and tells the user to type in the value of the principal. The **INPUT** statement on line 3Ø allows the user to feed in data at the keyboard when the computer requests it by printing a question mark. The letter P is called a **simple variable** and can be thought of as the name of a **store or location** inside the computer memory. Thus the effect of the INPUT command is the printing of the question mark, allowing the user to type in the principal (200 in this example), and the placing of this number in store P. Similarly line 4Ø tells the user to input the rate of interest, its value being placed in location R.

The actual calculation of the interest can now be performed because the data values are now in computer memory. In BASIC, the asterisk * is used for multiplication and the slant / for division. Line 6Ø is called an **assignment statement** and means multiply together the contents of P and R, then divide by 100 and place the answer in store I. The program must contain an instruction to print the interest, now in I. This is achieved by the PRINT command in line 7Ø which means print the contents of store I. Note the absence of quotation marks in line 7Ø:
PRINT "I" would print the letter I and not the value of the interest. The interest of 14 in this example has been printed on the last line of the computer run. The final line of the program contains an **END** statement indicating the end of the coding.

The ideas developed in this section will now be considered in greater detail.

NUMBERS AND THE E FORMAT

Numbers in BASIC may be positive, negative or zero. It will be assumed throughout this book that they may contain up to six significant figures and are in the range 10^{-616} to 10^{616}, although these restrictions do vary from system to system. Thus, −76.3, 124783, 0.00005, −.062 and 0 are legal numbers in BASIC. However, 176253.2 will be interpreted as 176253 because it has seven significant figures.

For very large or very small numbers it is necessary to use the **E format** as follows:
3.2E5 represents $3.2 \times 10^5 = 3.2 \times 100{,}000 = 320{,}000$
7.29E−6 represents $7.29 \times 10^{-6} = 7.29 \div 1{,}000{,}000 =$ 0.000 007 29.

ARITHMETIC OPERATORS

The operators +, −, *, / and ** or ↑ are used to represent addition, subtraction, multiplication, division and **exponentiation** (raising to a power), respectively.
For example:

A*B means A multiplied by B and must not be written in BASIC as AB, otherwise the computer will signify an error and not run the program until the error has been corrected.
A**2 or A↑2 both mean A squared and could also be written as A*A.
A**0.5 or A↑0.5 both mean the square root of A.
8/2 means 8 divided by 2 and will give the answer 4.

The order in which these operators are performed in complex expressions (not containing brackets) is determined by the following **hierarchy**:

Exponentiations are performed first.
Multiplications and divisions are performed second.
Additions and subtractions are performed third.

For example:

3+2*4 will give the answer 11 (multiplication before addition).
3+2*4**2 means 3+2*16, i.e. 35 (exponentiation first).

When the above hierarchy does not distinguish between the order of operations, they are performed from left to right: 8*4/2/5 means $(8 \times 4) \div (2 \times 5)$ in mathematical notation, i.e. 8 multiplied by 4 first, then divided by 2, then divided by 5.
11−8+4 gives the answer $3 + 4 = 7$ and not $11 - 12 = -1$.

Brackets may be used when it is required to perform operations in an order different to that entailed by the above conventions. Brackets are always performed first:

3*(2+4) will give 3*6 = 18.
(18+12)/(2*5) is 30 divided by 10, i.e. 3.
(A+B)/(3*C) is equivalent to (A+B)/3/C or in algebraic notation

$$\frac{A + B}{3C}$$

(A+B)*C/3 is equivalent to (A+B)/3*C or in algebraic notation

$$\frac{(A + B)C}{3}$$

It can be seen that line 60 of the program on page 11 could have been written in any of the following forms

```
60  I = P/100*R
60  I = 1/100*P*R
```

Clearly, the original line is by far the easiest to understand and illustrates a principle which should be followed – ensure that all programs are written in the simplest possible form in order that they may easily be read and understood.

SIMPLE VARIABLES

In standard BASIC a simple variable is either an alphabetic character A to Z or a single alphabetic character followed by a digit between 0 and 9 inclusive, e.g.

J, O, D8, A1, K9 are legal variables.
YB, S23, 7K, SUM are not allowed – the first contains two letters, the second two digits, the third has a digit first and the fourth contains three letters.

As noted on page 12 these variables can be thought of as the names of locations in which numbers can be placed. The simple interest program contains three such variables, i.e. P, R and I. Note that the choice of letters to represent each quantity gives the program clarity – consider the appearance of this program using Z for principal, O for rate of interest and VØ for interest. Of course, it will not always be possible to use the first letter of each quantity for its corresponding variable but a sensible choice of letters can still be made.

Note that in many BASICs this two-character restriction is not made. You should consult your manual to find out the restrictions on variable names for your computer.

ASSIGNMENT AND LET STATEMENTS

An assignment statement is a command which determines the contents of a store:

A=2 means place the number 2 in location A.
A=B*C means multiply the contents of locations B and C and place the answer in location A. If B and C contain 6 and 9 respectively, then this statement assigns the value 54 to the variable A.
S1=B2**2 means place the value of the square of the contents of location B2 in S1. If B2 contains 5 then the number 25 is placed in the location S1.
I=P*R/1ØØ, which is the determination of the interest in the program on page 11.
Note B*C=A, 2=A are illegal statements – the location for the answer must be placed first.

In these examples the statements are equivalent to the equations of algebra but the next assignment statement illustrates a fundamental use which has no mathematical correspondence:
X=X+1 means add 1 to the contents of X and place the answer back into X. In such a case the original contents of the location is lost. If the contents of X are 5 initially, then they are changed to 6 once this statement is executed and to 7 if the statement is executed once again.

Special care should be taken when using the same variable name to represent different quantities. To keep from becoming confused, a list of each variable used in the program with its intended use should be kept.

All assignment statements may be prefixed by the word **LET** as in the following:

```
LET A=(B+C)/2
LET I=P*T*R/1ØØ
```

The **LET statement** is equivalent to the previous assignment statements in this section but is compulsory in some versions of BASIC.

PRINT STATEMENTS

In this section the various possible forms of PRINT commands will be considered; page 30 shows its use with the TAB function. The PRINT command is an output statement (the only one in BASIC in fact) and is of fundamental importance.

PRINT "A" will print the letter A on the screen. The quotation marks ensure that the text in between them is output exactly as shown in the command.
PRINT "A=2" will print the equation A=2 on the screen.
PRINT " "HELLO" " will print "HELLO". The outer quotation marks are part of the BASIC language as noted above but the inner ones are punctuation symbols in standard English.

Further examples of this usage have already been noted on page 11. Note that blank spaces are reproduced exactly, as in the command:
PRINT" FRED BLOGGS " prints the name with a blank at the beginning and in the middle. The last blank is reproduced but, in this example, will not be discerned on the output.

In order to print the contents of a store, the quotation marks are omitted:
PRINT F outputs the contents of store F.
PRINT S+T will add the contents of S and T and then print the answer. This latter number will not be part of memory because of the lack of an assignment statement here.
PRINT 5 or PRINT 26−21 will both print the number 5 on the screen. Numbers do not necessarily have to be enclosed inside quotation marks, i.e. PRINT "5" does the same job but PRINT "26−21" outputs 26−21 (see the first example in this section).

Two or more numbers can be printed on the same line using the comma or semicolon, e.g. if A = 123 and B = 456,
PRINT A,B would produce 123   456
PRINT A;B would produce 123 456

The comma has left twelve spaces between the numbers, and the semicolon has left two spaces. The actual spacing obtained from these two symbols varies widely from computer to computer.
PRINT "A";"B" would produce AB. The letters have been run together because only expressions without quotes are automatically preceded and followed by a blank space (see also page 71).

The description of an answer and its value can now be combined on the same line.

PRINT"INTEREST IS";I gives INTEREST IS 42 assuming this line becomes the new line 7Ø of the program on page 11. Here text and the value of a variable are printed together, combining the two basic types of PRINT.
PRINT "INTEREST=$";I gives INTEREST=$ 14.

The comma and semicolon may also be employed as an instruction to the computer to remain on the same line after any data has been output (see also the next section).

```
 7Ø PRINT "A";
 8Ø PRINT "B";
 9Ø PRINT "C"
1ØØ PRINT "D";
```

The output of these four lines of code will be:

```
ABC
D
```

Line 7Ø prints the A, and the semicolon at the end ensures that a carriage return is not made onto the next line. Line 8Ø has a similar effect but line 9Ø prints the letter C and a carriage return is made to the next line because of the absence of a semicolon at the end of line 9Ø. The letter D is then printed on this new line; any further outputs also appear on this line, until another carriage return is executed.

Finally, the command PRINT by itself means print a blank space and make a carriage return.
PRINT "A" followed by PRINT leaves a line space after the A, i.e. print A, carriage return, print a blank (in other words, print nothing), carriage return.
PRINT "A"; followed by PRINT moves onto the line following the letter A, i.e. print A, no carriage return, print a blank, carriage return.

In Chapter 12 a version of the PRINT statement is also used to write data into a file.

INPUT STATEMENTS

Compare the program on page 11 with the following:

```
1Ø PRINT "A PROGRAM TO CALCULATE SIMPLE INTEREST"
2Ø P=2ØØ
3Ø R=7
4Ø I=P*R/1ØØ
5Ø PRINT "THE INTEREST IS $";I
6Ø END
```

This performs exactly the same calculation as the sample run on page 11 but there is a large difference between the two programs. The one above has the values within the program itself and contains no INPUT commands. Consequently, it will only perform one calculation when P = 200, R = 7 whereas the original can be run again with a different set of inputs. The INPUT statement is a very powerful tool – it allows the user to type in data while the program is running, thus giving generality to the program. Further input statements will be considered in Chapters 4 and 12.

Further variations of the INPUT command are listed below.
INPUT A means place the number input by the user into store A (see page 12).
INPUT A,B means place the two numbers input into stores A and B respectively. The user must type in the numbers with a comma separating them. Thus an input of 15,000 will not be interpreted as fifteen thousand but as two inputs of 15 and 000 (zero). The comma must not be used as a punctuation symbol in large numbers.
PRINT "PRINCIPAL";
INPUT P produces PRINCIPAL ?
i.e. the data is input on the same line because of the semicolon. Similarly this symbol (and the comma) can be placed at the end of an INPUT command as an instruction to remain on the same line.

Note that lines 20 and 40 of the first simple interest program may be omitted. In this case the question mark still appears on the screen but without any indication of its meaning. The user would not know that the value of the principal was required first, without inspecting the program.

The input of words to the computer is discussed in Chapter 9.

END STATEMENT

The word END is used to terminate all programs and is a signal to the computer that the program has been completed. In fact, in some computer systems it is optional, but it will be used throughout this book. The STOP statement (page 28) is another method of terminating a program, often without execution of some of the lines, but even in this case the last line of the program is the END statement.

MULTIPLE STATEMENTS

In some versions of BASIC, two or more commands can be placed on the same line as long as they are separated by the **backslash symbol** (\).

```
20  PRINT "PRINCIPAL"; \ INPUT P
```

has exactly the same meaning as

```
20  PRINT "PRINCIPAL";
30  INPUT P
```

The fact that INPUT P is on the same line as the PRINT command does not imply that the number is also input after the word PRINCIPAL – the semicolon achieves this. The maximum number of characters (including blank spaces) per line is usually 72 or thereabouts so that there is a limit to the number of instructions per line. It is advisable to use this symbol sparingly so that programs can be read without undue difficulty. It will be used in this book in order to save space.

Note The colon is sometimes used instead of the backslash.

COMPOUND INTEREST

Many business calculations involve compound interest in which the arithmetic is difficult to do manually. By using a computer these problems can be handled quickly and easily, removing the need for special financial tables.

```
10  PRINT "COMPOUND INTEREST"
20  PRINT "INITIAL AMOUNT"; \ INPUT P
30  PRINT "ANNUAL INTEREST RATE"; \ INPUT R
40  PRINT "NO. YEARS AND PERIODS PER YEAR"\INPUT Y,T
50  N=T*Y
60  R1=R/100/T
70  A=P* (1+R1) ↑N
80  I=A-P
90  PRINT "FINAL AMOUNT AND INTEREST"
100 PRINT A,I
110 END
```

Sample RUN to execute program

```
COMPOUND INTEREST
INITIAL AMOUNT ? 100
ANNUAL INTEREST RATE ? 12
NO. YEARS AND PERIODS PER YEAR
? 2,4
FINAL AMOUNT AND INTEREST
126.677                    26.677
```

The user types in the values of the variables P,R,Y,T, i.e. 100,12,2,4 – everything else is printed by the computer. Note the absence of the semicolon on line 40 and the input of Y and T. Line 50 calculates the total number of interest periods N by multiplying the number of years Y by the number of interest periods per year T. Line 60 calculates the interest rate per period R1 – in this run it represents the interest rate per quarter. The final amount A is then evaluated on the next line using the standard compound interest formula and the values of N and R1 evaluated on the previous two lines. The formula has been broken down into three formulae to avoid the complicated expression:
`70 A=P*(1+R/100/T) ↑ (T*Y)`

Finally, the actual interest I is evaluated on line 80 before the answers are printed.

EXCHANGE AND DUMP VARIABLES

A very common and important programming technique is to swap the contents of two variables. This would be needed for sorting data into order to determine the median or for putting a list of names into alphabetical order. The next two programs are attempts to achieve this.

```
10 INPUT A,B
20 A = B
30 B = A
40 PRINT A,B
50 END
```

```
10 INPUT A,B
20 D = A
30 A = B
40 B = D
50 PRINT A,B
60 END
```

Consider the first program with an input of 2 and 8 on line 10 – then location A contains 2 and location B contains 8. Line 20 means place the contents of location B (i.e. 8) into location A. Here the contents of B are unchanged – in other words the number 8 is not physically taken out of B but rather a "copy" of its contents. On the other hand the contents of A are changed from 2 to 8. The table below shows this:

Line number	A	B	Meaning
10	2	8	Input the numbers
20	8	8	Place contents of B in location A
30	8	8	Place contents of A in location B

The number 2 has been lost, both variables containing the number 8. The second program produces this table:

Line number	A	B	D	Meaning
10	2	8	0	Input the numbers
20	2	8	2	Place contents of A in location D
30	8	8	2	Place contents of B in location A
40	8	2	2	Place contents of D in location B

The initial contents of location D may be taken as zero before the assignment statement on line 20. The number 2 has been placed in this location first so that it is not lost.

D is called a **dump variable** because the contents of A are "dumped" into it to preserve their value. **Dumps** are rarely ever output but are, nevertheless, essential for the correct execution of this exchange procedure.

The two tables shown on this page are examples of a **dry run** or **desk run**, i.e. not the actual computer run of the program but one performed by hand by the programmer to check that the program does what it is intended to do.

Chapter 3

Control Statements

Most of the programs in Chapter 2 involve the evaluation of simple algebraic formulae, a task which can be equally well performed on a pocket calculator. In order to bring out the full potential of a computer, we need more programming functions which allow us to code complex techniques and algorithms. For example, to calculate the mean and standard deviation of one hundred numbers by the methods discussed so far would involve an extremely long and time-wasting program. In this chapter, commands which vary the order of line execution will be introduced.

RELATION OPERATORS

These are =, <, >, <=, >=, <> and are used to compare the value of one variable with another.

=   means is equal to
<   means is less than
>   means is greater than
<= means is less than or equal to
>= means is greater than or equal to
<> means is not equal to

These operators are always used with the **IF-THEN statement** which has the following form:
30 IF A = 7 THEN 60. This means if the contents of the variable A are equal to 7 then jump to line 60 of the program and if A is not equal to 7 then continue with the next line, i.e. line 40 in this example. Thus we can transfer to some other part of a program if a particular relationship is satisfied. In some versions of BASIC the control statement **IF-GOTO** is used in exactly the same way as IF-THEN.

Transfer of control may also be achieved by a **GOTO command** which has the following form:
70 GOTO 30
This means jump back to line 30 of the program irrespective of any particular condition being satisfied. For this reason it is called an **unconditional transfer statement** as opposed to the IF-THEN command which is a **conditional transfer statement**.

As an illustration, consider a program to calculate the traffic intensity and average number of customers the system contains for the simple single-server queue situation using the formulae:

$$\text{Traffic intensity } \rho = \frac{\text{Average arrival rate}}{\text{Average service rate}}$$

and average number of customers = $\rho/(1-\rho)$. The traffic intensity must be less than 1 for the latter formula to hold, otherwise the queue increases without limit.

```
 10  PRINT "ARRIVAL RATE";\ INPUT L
 20  PRINT "SERVICE RATE"; \ INPUT M
 30  T=L/M
 40  PRINT "TRAFFIC INTENSITY=";T
 50  IF T>=1 THEN 90
 60  A=T/(1-T)
 70  PRINT "AVERAGE NO. CUSTOMERS IN SYSTEM=";A
 80  GOTO 100
 90  PRINT "QUEUE INCREASES WITHOUT LIMIT"
100  END
```

Line 30 calculates the traffic intensity and line 40 prints the answer. On line 50 a check is made to see whether the value of T is greater than or equal to 1. If it is then a transfer is made to line 90 (missing out lines 60, 70 and 80) and the computer prints out the fact that the average is infinite and then ends. If the traffic intensity is not greater than or equal to 1, i.e. if it is less than 1, the program continues with line 60 to calculate A, the average number of customers in the system, prints the answer and then transfers unconditionally to line 100 skipping line 90. This GOTO statement is necessary to avoid printing out the text on line 90 in the case when A does have a finite value.

Statements such as IF-THEN and GOTO which allow the facility of jumping from one part of a program to another are called **control statements, transfer commands, branch commands** or **jump instructions**. The use of statements such as GOTO B and IF X<7 THEN B, where B is a variable, are illegal in the standard BASIC language–the line number must be explicit in the command.

SUMMATION

Statistical techniques are a major part of many business methods and it is important to be able to evaluate such measures as the arithmetic mean, standard deviation and correlation. These all involve the summation of a large number of quantities and it would be impractical to use a separate input statement for each one. In the case of a program requiring a variable amount of data, a method will be needed to inform the computer that all the data has been input. This can be achieved by the use of a **rogue value** (also known as **false data**) which is input at the end of the list. The rogue value has to be some number which could never be a part of the real data, e.g. −1 or 999, depending upon the situation. The program will check to see when this value is input by using an IF-THEN statement and then proceed to calculate the mean. The following variables are required:

N	the number of numbers input so far
S	the sum of the numbers input so far
A	the number currently being input
M	the arithmetic mean
−1	the rogue value

The program will perform the calculations as illustrated by this dry run.
Find the mean of the numbers 3, 7, 6, 8, 6.

A	**N**	**S**			
0	0	0	Initial values before data input		
3	1	3	First no. is 3,	increase N by 1,	add 3 to S
7	2	10	Next no. is 7,	increase N by 1,	add 7 to S
6	3	16	Next no. is 6,	increase N by 1,	add 6 to S
8	4	24	Next no. is 8,	increase N by 1,	add 8 to S
6	5	30	Next no. is 6,	increase N by 1,	add 6 to S
−1			Rogue value −	mean is $30 \div 5 = 6$	

S is calculated by **cumulative summation**. In each line the new value of S is the old value of S plus the value of A. The assignment statement to achieve this is S=S+A which means add S to A and place the answer in S. N increases by 1 every time a new number is input, corresponding to the statement N=N+1.

Program

```
5  N= 0 \S = 0
10 INPUT A
20 IF A = - 1 THEN 60
30 N = N + 1
40 S = S + A
50 GOTO 10
60 M = S/N
70 PRINT "MEAN=";M
80 END
```

Sample run

```
? 3
? 7
? 6
? 8
? 6
? -1
MEAN=6
```

Line 5 sets the values of N and S to zero initially. Although this line is not absolutely necessary, it is considered to be good programming practice to give **summation variables** their initial values by the use of assignment statements.

Line 20 checks each input – if it is the rogue value the program transfers to line 60 to calculate the mean, and line 70 prints the answer. If not, N and S are altered on lines 30, 40 and the GOTO statement returns to line 10 to input the next number.

Notice that the value of A is changed on each return to line 10 and that the five data values illustrated in this example are no longer stored in the computer at the end of the run (A contains the rogue value in the last line of the dry run). Although the sample run is somewhat artificial because of the small number of quantities input, the program would cope just as easily with a thousand numbers or more.

Lines 10 to 50 of the program form a type of **loop** because the order of line execution is 10, 20, 30, 40, 50, 10, 20, 30, 40, 50, 10, etc., until the condition on line 20 is satisfied and the program jumps out of the loop to line 60.

STANDARD DEVIATION

As an alternative to the use of rogue values, another way of informing the computer of the end of a data list is to input the number of quantities on the list. The next program calculates the mean and standard deviation of a set of numbers. The following formula is used:

$$\bar{x} = \frac{\sum fx}{n} \qquad s = \sqrt{\frac{1}{n}\sum fx^2 - \bar{x}^2}$$

where $n = \sum f$

Variables

L	number of different x values to be input
C	number of different x values so far
N	sum of frequencies so far
S1	sum of frequency multiplied by the x value
S2	sum of frequency multiplied by the x value squared
X	data value
F	frequency
M	arithmetic mean
S	standard deviation

```
 10 PRINT "MEAN AND STANDARD DEVIATION"
 20 PRINT "INPUT NUMBER OF DIFFERENT X VALUES"
 30 INPUT L
 40 C=0\N=0\S1=0\S2=0
 50 PRINT "INPUT X VALUES AND FREQUENCY"
 60 INPUT X,F
 70 C=C+1
 80 N=N+F
 90 S1=S1+F*X
100 S2=S2+F*X*X
110 IF C<>L THEN 60
120 M=S1/N
130 S=SQR(S2/N-M*M)
140 PRINT "MEAN=";M;"STANDARD DEVIATION=";S
150 END
```

Sample run

```
MEAN AND STANDARD DEVIATION
INPUT NUMBER OF DIFFERENT X VALUES
? 5
INPUT X VALUES AND FREQUENCY
? 0, 1
? 1, 4
? 2, 6
? 3, 4
? 4, 1
MEAN= 2   STANDARD DEVIATION= 1
```

The following dry run illustrates the values of the variables during the execution of the program.

L X F C	N	S1	S2
5 0 0 0	0	0	0
5 0 1 1	0 + 1 = 1	0 + 0 = 0	0 + 0 = 0
5 1 4 2	1 + 4 = 5	0 + 4 = 4	0 + 4 = 4
5 2 6 3	5 + 6 = 11	4 + 12 = 16	4 + 24 = 28
5 3 4 4	11 + 4 = 15	16 + 12 = 28	28 + 36 = 64
5 4 1 5	15 + 1 = 16	28 + 4 = 32	64 + 16 = 80

C now equal to L – transfer to line 120
M = S1/N = 32/16 = 2
S = SQR(S2/N−M*M) = SQR(80/16 − 2x2) = SQR(1) = 1

C is called a **counter** – when it reaches the same value as L all data have been input. Lines 60 to 110 form the loop which is executed as long as C is not equal to L. As soon as C becomes equal to L, the program exits from this loop to line 120 which then calculates the mean. Line 130 uses the **square root function SQR** to find the standard deviation. Note that the expression to be rooted must be placed in brackets after the letters SQR. Line 140 illustrates the printing of values of a variable in between sections of text using the semicolon.

MAXIMUM AND MINIMUM VALUES

The next program will find the maximum and minimum of a set of data. In practice this would only be part of a program in which the data concerned are calculated by the program itself, e.g. profit maximization.

Variables required:

L	largest value so far–set small initially
S	smallest value so far–set large initially
X	current data value
1111	rogue value

```
10  L=-1000\S=1000
20  INPUT X
30  IF X=1111 THEN 90
40  IF X<=L THEN 60
50  L=X
60  IF X>=S THEN 20
70  S=X
80  GOTO 20
90  R=L-S
100 PRINT "MINIMUM IS";S
110 PRINT "MAXIMUM IS";L
120 PRINT "RANGE IS";R
130 END
```

Here it is assumed that the data values lie somewhere in the range −1000 to 1000. On line 40, if X is less than or equal to L, the largest so far, transfer moves to line 60 and L does not change in value. If X is greater than L. line 50 is executed and L changes to the current value of X. L is given a low value initially so that the first X value is certain to be larger than it. Similar considerations apply to the minimum value S.

Dry run

X	L	S	X<=L	L	X>=S	S
5	−1000	+1000	No	5	No	5
7	5	5	No	7	Yes	5
2	7	5	Yes	7	No	2
12	7	2	No	12	Yes	2
6	12	2	Yes	12	Yes	2
1	12	2	Yes	12	No	1
1111	12	1	Stop, largest 12, smallest 1.			

L does not change when X<=L.
S does not change when X>=S.

DEBENTURES

Consider the problem of borrowing a given amount of money to be paid back over a certain period of years at a given rate of interest per annum. R represents the rate, A the amount borrowed, N the period and P the repayment at the end of each year. Then P is given by the formula:

$$P = \frac{AS^N (S - 1)}{(S^N - 1)}$$

$$\text{where} \quad S = 1 + \frac{R}{100}$$

The following additional variables will be used:

Y the current year
A the amount owing at the beginning of each year
B the amount owing at the end of each year
I the repayment on the interest each year
D the repayment on the original debt each year

```
10 PRINT "DEBENTURES"
20 PRINT "CAPITAL BORROWED";\ INPUT A
30 PRINT "RATE OF INTEREST";\ INPUT R
40 PRINT "REPAYMENT PERIOD";\ INPUT N
50 S=1+R/100
60 P=A*S↑N*(S−1)/(S↑N−1)
70 PRINT "YEARLY REPAYMENTS $";P
80 PRINT \ PRINT
90 Y= 0
100 PRINT "Y      A       I       D       P       B"
110 Y = Y + 1
120 I=A*R/100
130 D=P−I
140 B=A−D
150 PRINT Y;A;I;D;P;B
160 A=B
170 IF Y<>N THEN 110
180 PRINT\PRINT
190 P=P*N
200 PRINT "TOTAL REPAYMENT $";P
210 END
```

Sample run

```
DEBENTURES
CAPITAL BORROWED ? 50000
RATE OF INTEREST ? 14
REPAYMENT PERIOD ? 10
YEARLY REPAYMENT $ 9585.67


Y    A        I        D        P        B
1  50000    7000    2585.67  9585.67  47414.3
2  47414.3  6638    2947.66  9585.67  44466.6
3  44466.6  6225.33 3360.34  9585.67  41106.3
4  41106.3  5754.88 3830.79  9585.67  37275.5
5  37275.5  5218.57 4367.09  9585.67  32908.4
6  32908.4  4607.18 4978.49  9585.67  27929.9
7  27929.9  3910.19 5675.48  9585.67  22254.4
8  22254.4  3115.62 6470.04  9585.67  15784.4
9  15784.4  2209.82 7375.85  9585.67  8408.56
10 8408.56  1177.2  8408.47  9585.67  0.0878906


TOTAL REPAYMENT  $ 95856.7
```

The remaining debt at the end of the tenth year should be exactly zero but an error of approximately nine cents has been produced by the computer because it rounds numbers to six significant figures at each stage. Lines 80 and 180 have the effect of leaving two lines blank. Line 100 prints the heading for the table leaving the required number of blank spaces to center the letters as accurately as possible over their relevant columns. The computer has not produced perfect spacing between the columns because of the varying number of digits in each quantity printed. A neater method will be introduced in the next chapter. Lines 110 to 150 perform the relevant calculations for each year. On line 160 the value of the debt for the beginning of the next year is set equal to the debt at the end of the current year before looping back to line 110 (as long as Y is not equal to N).

PRESENT VALUES

A company is considering an investment in new machinery for which the net cash flows are as follows:

Year 0	Year 1	Year 2	Year 3
−4000	1200	3600	5400

Calculate the net present value (NPV) at a discount rate of 50% using conventional year end cash flows. To program this type of problem C will represent the cash flow in a particular year Y, P the NPV calculated cumulatively, D the discount factor and R the discount rate. A typical dry run for the program follows using the standard formula:
P is set to zero initially and D set to 1 because this is the factor for year 0.

Y	D	C	C/D	P
–	1	–	–	0
0	1	−4000	−4000	−4000
1	3/2	1200	800	−3200
2	9/4	3600	1600	−1600
3	27/8	5400	1600	0

Apart from the first row, the new value of D is always 3/2 multiplied by its old value corresponding to the equation D=D*3/2, i.e. D=D*(1+R/100) in general. The new value of P is calculated by adding C/D to its old value.

```
10  PRINT "PRESENT VALUES"
20  PRINT "DISCOUNT RATE";\INPUT R
30  Y=0\D=1\P=0
40  PRINT "INPUT THE CASH FLOWS"
50  PRINT "TYPE -1 TO END THE LIST"
60  PRINT "YEAR"; Y;\INPUT C
70  IF C=-1 THEN 120
80  P=P+C/D
90  D=D*(1+R/100)
100 Y=Y+1
110 GOTO 60
120 PRINT "NPV IS $";P
130 PRINT "TYPE 1 FOR RERUN WITH NEW DISCOUNT RATE"
140 PRINT "TYPE 2 FOR RERUN WITH SAME DISCOUNT RATE"
150 PRINT "TYPE 3 TO END THE RUN"
160 INPUT Q
170 IF Q=1 THEN 20
180 IF Q=2 THEN 30
190 END
```

Sample run

```
PRESENT VALUES
DISCOUNT RATE ? 10
INPUT THE CASH FLOWS
TYPE -1 TO END THE LIST
YEAR 0 ? -5000
YEAR 1 ? 1500
YEAR 2 ? 1400
YEAR 3 ? 1900
YEAR 4 ? 2000
YEAR 5 ? -1
NPV IS $ 314.187
TYPE 1 FOR RERUN WITH NEW DISCOUNT RATE
TYPE 2 FOR RERUN WITH SAME DISCOUNT RATE
TYPE 3 TO END THE RUN
? 1
DISCOUNT RATE ? 25
INPUT THE CASH FLOWS
TYPE -1 TO END THE LIST
YEAR 0 ? -2000
YEAR 1 ? 5400
YEAR 2 ? -3000
YEAR 3 ? -1
NPV IS $ 400
```

```
TYPE 1 FOR RERUN WITH NEW DISCOUNT RATE
TYPE 2 FOR RERUN WITH SAME DISCOUNT RATE
TYPE 3 TO END THE RUN
? 3
```

After the program has printed the net present value, an option of three possible alternatives is offered. If the input is 1, control passes to line 20 to feed in another discount rate for a new problem. An input of 2 returns the program to line 30 because the same discount rate is being used and is already stored in R. In both cases line 30 is executed to set Y, D and P to their initial values for the new run. The program is stopped by an input of 3 on line 160, but any number other than 1 or 2 will also have the same effect. In this example the 3-way branch requires only two IF-THEN statements because a check for Q=3 is unnecessary. When a program is to be run many times in succession a good deal of time can be saved by allowing the choice of a rerun within the program itself, but care must be taken to ensure that all variables are set to their correct values for any possible reruns.

INVENTORY CONTROL

The use of the **STOP statement** is illustrated by the next program which evaluates the Economic Order Quantity in one of two possible situations – either the basic case where replacement is instantaneous or the case with gradual replenishment.

```
10  PRINT "DEMAND PER ANNUM";\INPUT D
20  PRINT "ORDERING COST PER ORDER";\INPUT C1
30  PRINT "CARRYING COST PER ITEM PER ANNUM";\INPUT C2
40  PRINT "TYPE 1 FOR BASIC CASE – 2 FOR GRADUAL"
50  PRINT "REPLENISHMENT"; \ INPUT Q
60  IF Q=2 THEN 100
70  E=SQR(2*C1*D/C2)
80  PRINT "BASIC EOQ IS";E;"ITEMS"
90  STOP
100 PRINT "REPLENISHMENT RATE (PER ANNUM)"; \ INPUT R
110 E=SQR(2*C1*D/C2/(1-D/R))
120 PRINT "EOQ WITH GRADUAL REPLENISHMENT IS";E;"ITEMS"
130 END
```

When Q is 1, lines 70 and 80 are executed first – line 90 has the effect of terminating the program at some point other than the END statement. A program may contain many such STOP statements but its last line must always be the END statement.

Chapter 4

Input and Output Statements

DATA STATEMENTS

In Chapter 2 we saw how data could be entered into the computer during the running of a program by means of the INPUT statement. It is also possible to place data within the program itself by the use of the **READ-DATA** statement. The next program calculates an index of the increase of total production of a particular firm between the years 1975 and 1980. Three products A, B and C are manufactured and the following information is given:

	1975 Production	1975 Cost per unit	1980 Production
A	1700	2	2400
B	4000	1	2500
C	1300	3	5000

The 1975 production and costs will be entered as part of the program and the 1980 production input in the usual way.

```
10  X=Ø\S=Ø\T=Ø
20  READ P,C
30  W=P*C
40  S=S+W
50  INPUT P1
60  Q=P1/P*1ØØ
70  T=T+Q*W
80  X=X+1
90  IF X<>3 THEN 2Ø
100 I=T/S
110 PRINT "INDEX IS";I
120 DATA 17ØØ, 2, 4ØØØ, 1, 13ØØ, 3
130 END
```

Line 2Ø means read the values of P (1975 production) and C (1975 cost). The computer will then automatically search through the program for the **DATA statement** on line 12Ø and place the first number on the list in store P and the second in store C. DATA statements are conventionally placed near the end of a program so that the data can be easily checked. For longer lists, several lines may be employed to store the data as long as each of these lines begins with the word DATA. W represents the weight for the first product and S calculates the sum of the weights by the cumulative method. On line 5Ø the 1980 production of product A is input and then the quantity relative Q is evaluated. T is used to find the sum of all the quantity relatives multiplied by the weights and X is a counter to determine whether all values have been input. On the return to line 2Ø for product B, the **READ statement** will assign the next two values on the data list to the variables P and C.

Dry run

P	C	W	S	P1	Q	Q* W	T	X
1700	2	3400	3400	2400	141	479000	479400	1
4000	1	4000	7400	2500	62	248000	727400	2
1300	3	3900	11300	5000	385	1501500	2228900	3

I = 2228900/11300 = 197
(All calculations are to the nearest whole number.)

This program could be rerun with the production for another year in place of that for 1980 as long as 1975 is still taken as the base year. The READ-DATA statement is used in place of the INPUT statement when data, which does not change in value, is required for a program. It avoids the need to input part of the data on every occasion that the program is run. To alter the base year, the program can be revised by typing in a new line 12Ø.

Sometimes one may wish to reuse data within a program. This may be achieved by the **RESTORE statement** which allows the program to go back to the beginning of a data list. For example, the addition of the line 25 RESTORE to the program would result in the values of 1700 and 2 being read again on the next return to line 2Ø. Data lists can be longer than is necessary, i.e. not all the data may be read in a program, but any attempt to read beyond a data list (the list is too short) will result in an error. Programs may contain many READ statements but care must be taken to place the values in the DATA statements in the correct order for reading and to ensure that there are sufficient values to supply each of the variables in the READ statements.

THE TAB FUNCTION

This allows the programmer to decide the positions at which characters are displayed on a line of output. It is used in conjunction with the PRINT statement and has the form **TAB(X)** where X is the position at which the character is to be printed. It will be assumed that a line of output contains 72 positions. For example:

PRINT TAB(12);"*" will print an asterisk at position 12 counted from the beginning of the line.
PRINT TAB(23);A will print the value of the variable A at position 23.
PRINT TAB(6);"X";TAB(12); "Y" will print the letter X at position 6 and the letter Y at position 12 on the same line of the output.
PRINT TAB(3);A;TAB(1Ø);B;TAB(17);C; will print the values of the variables A, B and C at positions 3, 1Ø and 17 respectively and then remain on the same line for the next PRINT statement because of the semicolon at the very end of the command.
PRINT TAB(17);"A";TAB(13);"B" will print A at position 17 and B at position 18 because the cursor cannot move back along the line from position 17 to position 13 after the printing of the letter A.

PRINT TAB(A);"X" will print the letter X at position A where A is a variable whose value has been assigned by a previous statement.
PRINT TAB(12.2);"A" prints an A in position 12
PRINT TAB(32.8);"A" prints an A in position 32
i.e. the TAB positions are rounded down.
The TAB function is ideally suited for printing out tables or graphs.

TABLES

A mill has n machines, each of which has independently a probability, p, of breaking down. To evaluate the probability of r breakdowns we need the standard formula for the binomial distribution:

$$P(r)=\frac{n!}{r!(n-r)!}\,p^r q^{n-r} \quad \text{where} \quad q=1-p$$

To program this problem, it is simpler to transform this formula into the following:

$$P(r+1)=\frac{(n-r)p}{(r+1)q}\,P(r) \quad \text{where} \quad P(0)=q^n$$

This is called an **inductive or recursive formula** and gives the probability of r+1 breakdowns, P(r+1), in terms of the probability of r breakdowns, P(r).

Variables

N the number of machines
P the probability of a machine breaking down
Q the probability of a machine not breaking down
R the number of breakdowns (ranging from 0 to N)
A the probability of R breakdowns
C the cumulative probability

```
10  PRINT "NUMBER OF MACHINES";
20  INPUT N
30  PRINT "PROBABILITY OF A BREAKDOWN";
40  INPUT P
50  PRINT\PRINT
60  PRINT TAB(1);"NO. B/DOWNS";
70  PRINT TAB(16);"PROBABILITY";
80  PRINT TAB(31);"CUM. PROB."
90  PRINT
100 Q=1-P\R=0
110 A=Q↑N\C=A
120 PRINT TAB(5);R;TAB(18);A;TAB(32);C
130 IF R=N THEN 180
140 A=A*(N-R)*P/(R+1)/Q
150 C=C+A
160 R=R+1
170 GOTO 120
180 END
```

Line 50 prints two blank lines after the input, and lines 60 to 80 print the column headings starting in the positions stated. Note the use of the semicolon at the end of lines 60 and 70 and its absence in line 80. It would have been possible to compress these three lines into a single PRINT statement. The variables are initialized on lines 100 and 110 so that A represents the probability of zero breakdowns and C has this same value. Line 120 then prints these values under the corresponding headings. A check is then made to see if the table is complete – when R = N the previous line must have just printed the last row of the table. Otherwise the new value of A is evaluated using the inductive formula, C is incremented by this new value and R increased by one before looping back to the PRINT statement.

Sample run

```
NUMBER OF MACHINES ? 12
PROBABILITY OF A BREAKDOWN ? 0.3

NO. B/DOWNS        PROBABILITY        CUM. PROB.

0                  0.0138413          0.0138413
1                  0.0711837          0.085025
2                  0.16779            0.252815
3                  0.2397             0.492515
4                  0.231139           0.723654
5                  0.158495           0.88215
6                  0.0792477          0.961397
7                  0.0291114          0.990509
8                  0.00779769         0.998306
9                  0.00148527         0.999792
10                 0.00019096         0.999982
11                 0.00001488         0.999997
12                 .531438E-006       0.999998
```

The final probability in the table uses the E format and represents the number 0.00000531438. The last cumulative probability should be exactly one – an error of 0.000002 has been made by the computer.

GRAPHS

At this stage only graphs of given functions will be considered. The following equations represent the relationship between costs, production and revenue for a certain product:

Revenue R=x/3 Cost C=x/6+1.5

where x is the production, in thousands of units, and C and R are measured in thousands of dollars. The graph is plotted for values of x between 0 and 24, the x-axis appearing as the vertical axis. Thus each new line of the output will represent a different x value. The costs and revenues are plotted on the horizontal axis using the TAB function. A consideration of the two equations reveals that the minimum value of C or R occurs when x=0, in which case R=0. Similarly the maximum value is R=8 when x=24. A plot of this range of values would produce a very small graph with only TAB positions between 0 and 8 being used. To overcome this problem the values of C and R will be multiplied by 6 and then 5 added to them. R=0 will correspond to TAB position 5 and R=8 to position 53, the first four positions being used to print the x values.

```
10  PRINT TAB(5);"Ø";TAB(23);"REVENUE/COST";TAB(53);"8"
20  X=Ø
30  R=X/3
40  C=X/6+1.5
50  R=6*R+5
60  C=6*C+5
70  PRINT TAB(2);X;
80  IF R>C THEN 120
90  IF R<C THEN 140
100 PRINT TAB(R);"O"
110 GOTO 150
120 PRINT TAB(C);"*";TAB(R);"+"
130 GOTO 150
140 PRINT TAB(R);"+";TAB(C);"*"
150 IF X=24 THEN 180
160 X=X+1
170 GOTO 30
180 END
```

Computer run

```
     Ø                 REVENUE/COST                  8
  Ø  +        *
  1    +       *
  2      +      *
  3        +     *
  4          +    *
  5            +   *
  6              +  *
  7                + *
  8                  +*
  9                    0
  1Ø                    *+
  11                     * +
  12                      *  +
  13                       *   +
  14                        *    +
  15                         *     +
  16                          *      +
  17                           *       +
  18                            *        +
  19                             *         +
  2Ø                              *          +
  21                               *           +
  22                                *            +
  23                                 *             +
  24                                  *              +
```

Line 1Ø prints the revenue/cost axis with the scale in the appropriate positions. X is set to its initial value on the next line before the actual and scaled values of R and C are evaluated (lines 3Ø to 6Ø inclusive). Line 7Ø simply prints the x value at the beginning of each row in TAB position 2. An asterisk is used for the cost line and a plus sign for the revenue line. If the lines intersect at one of the x values in the program a letter O is printed instead – this is accomplished by line 1ØØ which is only executed when R=C. The condition on lines 8Ø and 9Ø is needed to ensure that the printing of the points on lines 12Ø and 14Ø is done correctly – the first TAB position in these PRINT statements must be smaller than the second (see the example at the bottom of page 30).

The program can be generalized in many ways. The range of x values and the step could be inputs to the computer in which case the scale would also have to be input or calculated by the program itself. The equations of the lines may be varied by inputting the coefficients of x and the constant terms, though not all lines will appear perfectly plotted because of the restriction of 72 positions on a row. This can be improved slightly by rounding off each TAB position to the nearest whole number using one of the special functions considered in Chapter 7. It is possible to draw the graph of any equation with two variables by means of the TAB function though this will be deferred to a later chapter when further BASIC statements are available.

Some versions of BASIC include powerful plotting statements which are able to draw the actual straight lines or curves as opposed to the TAB function which only prints the points without joining them up. Unfortunately there are a large number of such plotting routines which vary a good deal from system to system. These routines are not particularly difficult to program but will not be considered in this book.

Chapter 5

Loops

FOR-NEXT STATEMENT

When the exact number of loops to be executed in a program is known, BASIC provides a construction known as the **FOR-NEXT statement** which can be employed instead of the method of looping considered in Chapter 3. It has the form:

FOR I=A TO B STEP S

Here I is the variable which takes different values starting at A and increasing to B in steps of S on each pass through the loop. For example:

FOR I = 1 TO 7 STEP 2 I takes the values 1, 3, 5, 7
FOR I = 3 TO 18 STEP 5 I takes the values 3, 8, 13, 18
FOR I = 6 TO 2 STEP−1 I takes the values 6, 5, 4, 3, 2
FOR I = 1 TO 5 I takes the values 1, 2, 3, 4, 5
(If the step is omitted, it is assumed to be 1.)
FOR I = 2 TO 3 STEP Ø.4 I takes the values, 2, 2.4, 2.8 because the next value of 3.2 is larger than the finishing value of 3.
FOR I = 4 TO 3 STEP 2 I takes only the value 4

i.e. a FOR-NEXT loop always takes the starting value no matter what the step is.

The initial value, final value and step may also be variables as in FOR I=X TO Y STEP Z. In the above examples I is called the **loop variable** or the **control variable**. The letters I, J and K are commonly used to represent control variables though there is no compulsion to follow this convention. The end of a loop is terminated by the command **NEXT I** (assuming I is the loop variable).

As a very simple illustration the following program will print the squares and cubes of all numbers from 1 to 1ØØ.

```
1Ø  FOR I=1 TO 1ØØ
2Ø  S=I↑2
3Ø  C=I↑3
4Ø  PRINT I,S,C
5Ø  NEXT I
6Ø  END
```

Dry run I	S	C
1	1	1
2	4	8
3	9	27
4	16	64
ETC.	ETC.	ETC.
100	10000	1000000

I takes an initial value of 1 on the first line. The next two lines calculate the square and cube of I and line 4Ø prints the answer. On reaching line 5Ø, control returns to line 1Ø where I is incremented by 1 and becomes 2, the loop then being executed again. After 100 such passes, the program automatically jumps out of the loop to the line after the NEXT statement which is the end of the program in this example. It is essential to have a corresponding NEXT statement for each FOR statement.

MEAN AND MODE

The use of statements such as INPUT, GOTO and IF-THEN may be combined with FOR-NEXT loops as the next example shows.

```
 10  PRINT "MEAN AND MODE"
 20  PRINT "NO. OF DIFFERENT VALUES"
 30  INPUT M
 40  N=0\S=0\L=-1
 50  PRINT "INPUT VALUES AND FREQUENCY"
 60  FOR I=1 TO M
 70  INPUT X,F
 80  N=N+F
 90  S=S+F*X
100  IF F<L THEN 130
110  L=F
120  X1=X
130  NEXT I
140  A=S/N
150  PRINT "MEAN IS";A
160  PRINT "MODE IS";X1;"WITH FREQUENCY";L
170  END
```

Dry run with M=5

I	X	F	N	S	L	X1
—	—	—	0	0	−1	—
1	2	1	1	2	1	2
2	3	3	4	11	3	3
3	4	2	6	19	3	3
4	5	4	10	39	4	5
5	6	2	12	51	4	5

Mean A = S/N = 51/12 = 4.25
Mode X1 = 5 with frequency L = 4

The first line of the dry run represents line 40 of the program. N and S are used to calculate the sum of the frequencies and the sum of all the values respectively by the standard technique. L stores the largest frequency so far and X1 the corresponding data value. These two variables only change in value when the current frequency is greater than or equal to L. It is worth noting that the condition on line 100 passes control to the NEXT statement first when the frequency is less than L – every pass through a loop must end with the NEXT command. The control variable I is not needed in any of the calculations here and only takes the role of a counter. In the case of a frequency distribution with more than one mode, this program will only print the one that is input last of all.

Rogue values can also be used in conjunction with FOR-NEXT loops by changing line 60 to FOR I=1 TO 999999 or some other very large finishing value. Lines 20 and 30 can now be omitted but an extra line immediately after the

INPUT statement on line 70 is needed, e.g. 75 IF X=−1 THEN 140 taking −1 as the rogue value. The loop will therefore not be executed 999999 times because this new line allows exit from the loop at a point where the control variable does not have its final value. Jumping out of a loop in this way is vital in some applications, but jumping into a loop by the use of a control statement (this does not include transfer to the FOR statement) is bad programming practice, as the control variable will not necessarily have the value expected on entry or the operation itself may be classified as illegal on some computer systems. When a control variable's starting value, final value or step are also variables, any attempt to change these values within the loop itself by the use of assignment statements will often have no effect whatsoever on the values taken by the control variable. It is best to avoid this practice at all times because of its uncertainty. Note that it is still permissible to alter the starting value, final value and step outside the loop itself.

STOCK CONTROL

A company uses 60 000 components of a certain type in its production process. Each component costs $12 to purchase and the carrying costs are 20% per annum. The ordering and handling costs are $170 per order. It is required to print out a table of values to illustrate the total cost to the company of various order quantities.

```
10  PRINT "NO. OF COMPONENTS PER ANNUM"\INPUT N
20  PRINT "PURCHASING COST OF A COMPONENT"\INPUT P
30  PRINT "CARRYING COSTS (% PER ANNUM)"\INPUT C
40  PRINT "ORDERING AND HANDLING COSTS"\INPUT H
50  PRINT "STARTING VALUE, FINISHING VALUE AND STEP"
60  PRINT "FOR ORDER QUANTITIES TO BE CONSIDERED"
70  INPUT Q1, Q2, S
80  PRINT TAB(1);"ORDER",TAB(11);"NUMBER";TAB(21);"ORDER";
90  PRINT TAB(31);"MEAN";TAB(41);"HOLDING";TAB(51);"TOTAL"
100 PRINT TAB(1);"SIZE";TAB(11);"ORDERS";TAB(21);"COST";
110 PRINT TAB(31);"STOCK";TAB(41);"COST";TAB(51);"COST"
120 T1=1E12
130 FOR I=Q1 TO Q2 STEP S
140 O=N/I
150 C1=O*H
160 A=I/2
170 C2=A*P*C/100
180 T=C1+C2
190 PRINT TAB(1);I;TAB(11);O;TAB(21);C1;
200 PRINT TAB(31);A;TAB(41);C2;TAB(51);T
210 IF T>T1 THEN 230
220 T1=T\I1=I
230 NEXT I
240 PRINT\PRINT
250 PRINT "OPTIMUM ORDER QUANTITY IS"; I1
260 END
```

T1 represents the lowest total cost so far and I1 the corresponding order quantity. I is the loop variable used to represent the order quantity, O the average number of orders per annum, C1 the annual ordering and handling costs, A the average stock level, C2 the annual stock holding cost and T the total cost. The problem in this section will produce the following table of values using the inputs N=60000, P=12, C=20, H=170, Q1=1000, Q2=6000, S=1000.

ORDER SIZE	NUMBER ORDERS	ORDER COST	MEAN STOCK	HOLDING COST	TOTAL COST
1000	60	10200	500	1200	11400
2000	30	5100	1000	2400	7500
3000	20	3400	1500	3600	7000
4000	15	2550	2000	4800	7300
5000	12	2040	2500	6000	8040
6000	10	1700	3000	7200	8900

OPTIMUM ORDER QUANTITY IS 3000

MULTIPLE LOOPS

Loops may also be placed inside one another; this technique is called **multiple looping** or **nested looping**, i.e. one loop inside another or two separate loops inside a main loop. For example a **double loop**:

```
10 FOR I=1 TO 2
20 FOR J=1 TO 3
30 PRINT I,J
40 NEXT J
50 NEXT I
```

Output

I	J
1	1
1	2
1	3
2	1
2	2
2	3

The main loop from lines 10 to 50 contains the inner loop from lines 20 to 40. On the first pass through the outer loop, the inner loop is executed to produce the first three pairs of numbers in the output. On completion of the inner loop control passes to line 50 and the next value of I is assigned; the inner loop is executed once again.

The **triple loop** is commonly seen in practice and has the following form:

```
10 FOR I=1 TO 3
20 FOR J=1 TO 2
30 FOR K=1 TO 4
40 PRINT I, J, K
50 NEXT K
60 NEXT J
70 NEXT I
```

Dry run

I J K	I J K	I J K
1 1 1	2 1 1	3 1 1
1 1 2	2 1 2	3 1 2
1 1 3	2 1 3	3 1 3
1 1 4	2 1 4	3 1 4
1 2 1	2 2 1	3 2 1
1 2 2	2 2 2	3 2 2
1 2 3	2 2 3	3 2 3
1 2 4	2 2 4	3 2 4

The I loop is performed once, the J loop three times and the K loop six times altogether. The inner loop variable changes the fastest of all and the outer loop the slowest. Every FOR statement must have its corresponding NEXT statement. The order of the NEXT statements is crucial and must obey the following rule in order to avoid the error of jumping into the middle of a loop–all inner loops must be wholly contained within their outer loops. For example:

```
FOR I = 1 TO 5
FOR J = 1 TO 3
NEXT J
FOR K = 1 TO 4
NEXT K
NEXT I
```

```
FOR I = 1 TO 5
FOR J = 1 TO 3
FOR K = 1 TO 4
NEXT J
NEXT K
NEXT I
```

The first illustration is legal–the rule does not apply to the J and K loops because the J loop ends before the K loop starts, i.e. they are separate loops. The second example is illegal as the K loop is not completely contained within the J loop. The pictorial representation shown is perhaps easier to understand–the first example has no "crossing" loops, while the second has.
Larger numbers of nested loops are occasionally required and computer systems often allow as many as ten or more loops nesting one inside the other.

CASH FACTORS

The double loop is illustrated in the next program which will produce either compound interest factors or present value factors.

```
10 PRINT "TYPE 1 FOR PRESENT VALUE FACTORS"
20 PRINT "TYPE 2 FOR COMPOUND INTEREST FACTORS"
30 INPUT Z
40 PRINT "START, FINISH, STEP FOR NUMBER OF YEARS"
50 INPUT N1,N2,S
60 IF Z=2 THEN 90
70 PRINT "START AND STEP FOR DISCOUNT RATE"
80 GOTO 100
90 PRINT "START AND STEP FOR INTEREST RATE"
100 INPUT R1,T
110 R2=R1+6*T\Z=2*Z-3
120 PRINT TAB(1);"N/R";
130 P=7
140 FOR R=R1 TO R2 STEP T
150 PRINT TAB(P);R;
160 P=P+8
170 NEXT R
180 PRINT
190 FOR I=1 TO 59\PRINT "-";\NEXT I\PRINT
200 FOR N=N1 TO N2 STEP S
210 PRINT TAB(1);N;
220 P=5
230 FOR R=R1 TO R2 STEP T
240 F=(1+R/100)↑(Z*N)
250 PRINT TAB(P);F;
260 P=P+8
270 NEXT R
280 PRINT
290 NEXT N\END
```

R2 represents the final value for the rate and is designed to ensure that the rate takes exactly seven values, allowing each line of the table to fit neatly on the output. Also, on line 110, the value of Z is changed for a later part of the program – if Z=1 is input, Z becomes 2*1−3=−1 and if the input is Z=2, Z becomes 2*2−3=1. Line 120 prints a heading for the variables in the table – N stands for the number of years and R for the rate. The loop from 140 to 170 will print the values of the rate along the top row of the table using the variable P as the TAB position. On the first pass through this loop, the initial rate will be printed in position 7, P becomes 15 (line 160) and control then returns to the beginning of the loop. On the second pass the next rate will be printed in position 15 and P then changes to 23. In this manner the seven rates appear in positions, 7, 15, 23, 31, 39, 47, 55. After the completion of the loop the cursor will still be on the top row of the table because of the semicolon at the end of line 150. To move onto the next line, the PRINT statement on line 180 has been inserted; this literally means print a space at the end of the row and execute a carriage return. Line 190 underlines this row, i.e. it prints 59 dashes on the next row of the table.

The double loop from lines 200 to 290 calculates and prints the main part of the table. The number of years N is varied in the outer loop which begins by printing N in the first place of each row. P is again used to determine the TAB positions for the factors. It is also possible to work out a formula for P in terms of the inner loop variable R(P=8*(R−R1)/T+5) and to place this immediately in front of line 250, omitting lines 220, 260. For each value of R the factor F is evaluated – for compound interest Z=1 now and the power on line 240 is N, for present values Z=−1 and this power is −N, meaning take the Nth power and invert. The final PRINT statement on line 280 moves the cursor onto the next row of the table.

Sample run

```
TYPE 1 FOR PRESENT VALUE FACTORS
TYPE 2 FOR COMPOUND INTEREST FACTORS
? 2
START, FINISH, STEP FOR NUMBER OF YEARS
? 1, 8, 1,
START AND STEP FOR INTEREST RATE
? 11, 2
N/R  11        13        15        17        19        21        23
------------------------------------------------------------------------
1    1.11      1.13      1.15      1.17      1.19      1.21      1.23
2    1.2321    1.2769    1.3225    1.3689    1.4161    1.4641    1.5129
3    1.36763   1.4429    1.52087   1.60161   1.68516   1.77156   1.86087
4    1.51807   1.63047   1.749     1.87389   2.00534   2.14359   2.28886
5    1.68506   1.84243   2.01135   2.19244   2.38635   2.59374   2.8153
6    1.87041   2.08195   2.31306   2.56516   2.83976   3.13842   3.46282
7    2.07616   2.3526    2.66001   3.00124   3.37931   3.79749   4.25927
8    2.30453   2.65844   3.05902   3.51145   4.02138   4.59496   5.2389
```

In this run R1=11, T=2 so that R2=R1+6*T=11+6*2=23

and R ranges from 11 to 23 in steps of two giving a total of seven values for the rate. By adding a third loop around the N and R loops, it would be possible to print a series of tables for varying values of the initial principal or cash flow.

FORECASTING

The method of exponential smoothing is a very useful and simple technique to apply. It is based upon the formula:

$$\text{New forecast} = \alpha \times \text{latest observation} + (1 - \alpha) \times \text{old forecast}$$

Here the next prediction is a weighted average of the last prediction and the latest observation. α is the smoothing constant and takes a value between 0 and 1, although values between 0.1 and 0.3 are usually most suitable in practice. The program will predict the next month's sales figures of a company using varying values of the smoothing constant. The previous sales are placed in a data statement at the end of the program, though forecasts are also made for these to illustrate the accuracy of the method.

```
10  FOR A=0 TO 1 STEP 0.1
20  PRINT "SMOOTHING CONSTANT=";A
30  PRINT "MONTH";TAB(11);"SALES";TAB(25);"FORECAST"
40  F=0
50  FOR I=1 TO 13\PRINT TAB(3);I;
60  IF I=1 THEN 110\IF 1=2 THEN 100
70  F=A * S + (1-A) *F
80  IF I <> 13 THEN 110
90  S=0\GOTO 120
100 F=S
110 READ S
120 PRINT TAB (11);S;TAB(25);F
130 NEXT I
140 RESTORE\PRINT
150 NEXT A
160 DATA 35, 34, 36, 31, 28, 30, 27, 26, 31, 35, 37, 39
170 END
```

A represents the smoothing constant, F the monthly forecasts, I the number of the month and S the sales figures. When I is 1, the program jumps to line 110 to read the sales and then prints the forecast as zero, indicating that no data is available in the first month to make a prediction. When I is 2, line 100 makes the second month's forecast equal to the first month's sales as no old forecast is yet available. Otherwise, line 70 finds the new forecast by the standard formula. When I is 13, i.e. the month without a sales figure, S is set to zero, to indicate that sales are unknown, and line 110 is omitted, to avoid the error of reading beyond the end of a data list. After the completion of the I loop the data is restored in readiness for the next value of A.

Chapter 6

Lists

SUBSCRIPTED VARIABLES

In many applications it is necessary to input a very large number of variables into the computer and store them all in different locations, in contrast to earlier programs in this book where one variable was often adequate to store different values in succession. Such a situation may arise when the program is a large one and various operations have to be performed on the data. Variables such as A,B,C,D, . . . , or A1,A2,A3,A4, . . . , might be used but would make the program tedious to code because each of these simple variables would have to be mentioned explicitly, e.g. to write a statement to find the mean of eight variables would require the equation (A+B+C+D+E+F+G+H)/8. Fortunately BASIC provides an advanced type of variable called a **subscripted variable** which has the form A(1),A(2),A(3), . . . , or B(1),B(2), . . . , etc. The number in brackets is called the **subscript** and is preceded by any of the letters A to Z. C(13) means the thirteenth number on the C list, X(2) the second number on the X list. In fact it is permissible to use D(Ø) as a subscripted variable but expressions such as F(−1), H(1.5) and V(−2.7) are illegal – the subscript must be a non-negative integer. Note also that a subscripted variable cannot be used as a loop variable (see page 35).

These variables help to solve the problem because the set Y(1),Y(2),Y(3), . . . , Y(8) can be referred to generally as Y(I) where I ranges from 1 to 8, e.g. inside a loop. Any attempt to refer to the simple variables Y1,Y2, . . . , as YI would disobey the rules for simple variables. To input eight numbers each into a separate location can be achieved by:

```
10  FOR I=1 TO 8
20  INPUT A(I)
30  NEXT I
```

When I=1, the first number input will be placed in A(1), when I=2 the second number is placed in A(2) and so on. Inputting the values 3,5,Ø,7,6,1,5,8 would result in the list A(1)=3, A(2)=5, A(3)=Ø, A(4)=7, A(5)=6, A(6)=1, A(7)=5, A(8)=8.

Subscripted variables need a special instruction called a **dimension statement** to instruct the computer to reserve enough space for them in its memory. This statement has the form **DIM A(5Ø)** and has to be placed before any mention of the variable A(I) in the program – DIM statements are best written on the first line of a program.

DIM A(5Ø) will reserve space in the memory for 51 variables, i.e. A(Ø), A(1), . . . , A(5Ø).
DIM S(12), T(18) will reserve space for the 13 variables S(Ø), S(1), . . . , S(12) and the 19 variables T(Ø), T(1), . . . , T(18).

On some computers DIM statements are not needed if the list contains less than ten elements but all subscripted variables will be dimensioned in all further programs. On some computers DIM(N) (where N is a simple variable) is an illegal command. This is convenient if it is available since a program, for example, which requires 20 inputs on one run and 100 on another would not have to be dimensioned prior to running the program. You should check your manual to see if this is legal with your BASIC.

SORTING

It is now possible to write a program to sort data into numerical order using the concept of a dump variable which was introduced in Chapter 2. The **algorithm** (a finite set of rules to perform a given task) used here is known as the **bubble-sort technique**. To sort the numbers 5, 2, 3, 1 into order requires the following steps.

STEP 1 Compare the first number with the second and interchange if necessary.
This gives 2, 5, 3, 1.

STEP 2 Compare the first number with the third number (on the new list) – interchange if necessary.
This gives 2, 5, 3, 1 (no change required).

STEP 3 Compare the first number with the fourth number.
This gives 1, 5, 3, 2.

At this stage the smallest number will always be at the front of the list. Continuing:

STEP 4 Compare the second and third numbers.
This gives 1, 3, 5, 2.

STEP 5 Compare the second and fourth numbers.
This gives 1, 2, 5, 3.

Now the next smallest number is in its correct place.

STEP 6 Compare the third and fourth numbers.
This gives 1, 2, 3, 5.

Now all numbers are in the correct order.

Steps 1 – 3 compare the first number with the other three.
Steps 4 – 5 compare the second number with the ones after it.
Step 6 compares the third number with the one after it.
Let us call the position of this number I, i.e. I=1 in the first three steps, I=2 in the next two and I=3 in the last. If J is used to represent the position of the second number in each comparison we obtain the following:

Step 1	I=1 with J=2	Step 4	I=2 with J=3
Step 2	I=1 with J=3	Step 5	I=2 with J=4
Step 3	I=1 with J=4	Step 6	I=3 with J=4

For a given value of I, the J value ranges from one more than

I to 4, the number of values in the list. Letting N be the number of values in the general case, J ranges from I+1 to N. I itself ranges from 1 to 3 in this example and from 1 to N−1 in general. Thus we can use a double loop to perform all the comparisons. Four numbers require six comparisons, five numbers require ten comparisons, ten numbers require forty-five comparisons and N numbers require N*(N−1)/2 comparisons.

```
1Ø  DIM A(1ØØ)
2Ø  PRINT "SORTING APPLICATION"
3Ø  PRINT "NUMBER OF VALUES IN THE LIST"
4Ø  INPUT N
5Ø  PRINT "INPUT THE NUMBERS"
6Ø  FOR I=1 TO N
7Ø  INPUT A(I)
8Ø  NEXT I
9Ø  FOR I=1 TO N-1
1ØØ FOR J=I+1 TO N
11Ø IF A(I)<=A(J) THEN 15Ø
12Ø D=A(I)
13Ø A(I)=A(J)
14Ø A(J)=D
15Ø NEXT J
16Ø NEXT I
17Ø FOR I=1 TO N
18Ø PRINT A(I);
19Ø NEXT I
2ØØ END
```

Line 1Ø reserves space for 101 locations in memory, though A(Ø) will never be used in the program. I has been used in three different loops in the program but this will not lead to error because each loop ends before the next one starts. On line 11Ø, I and J represent the position of the numbers to be compared – A(I) and A(J) are the actual numbers. If A(I) is less than or equal to A(J) the two numbers must already be in the correct order (I is always less than J in each comparison) so transfer is made to the next value of J. If not, the numbers in the stores A(I) and A(J) are interchanged using the dump variable D. In the illustration shown A(1)=5, A(2)=2, A(3)=3, A(4)=1 and for the first comparison when I=1 and J=2:

line 11Ø becomes IF A(1)<=A(2) THEN 15Ø, i.e. IF 5<=2 THEN 15Ø which is not true so line 12Ø is executed.
line 12Ø becomes D=A(1), i.e. D=5 (put value of A(1) in D)
line 13Ø says A(1)=A(2), i.e. A(1)=2 (put value of A(2) into A(1))
line 14Ø says A(2)=D, i.e. A(2)=5 (put value of D into A(2))
Thus A(1)=2, A(2)=5, A(3)=3, A(4)=1 which is the order after step 1.

PROFIT MAXIMIZATION

A bakery sells a certain type of loaf of bread. Each loaf costs

13 cents to make, and is sold for 32 cents. Any loaves unsold at the end of the day are sent to a farm and bring in 6 cents each. The anticipated daily demand based on past sales is as follows:

Demand	Frequency	Demand	Frequency
70	3	110	19
80	11	120	6
90	22	130	2
100	37		

The problem is to determine the number of loaves made per day to give the maximum possible expected profit.

```
10  DIM D(20),F(20)
20  PRINT "COST OF MAKING PRODUCT";\INPUT C1
30  PRINT "NORMAL SELLING PRICE";\INPUT S1
40  PRINT "LOWER SELLING PRICE";\INPUT S2
50  F=0\M=-1E12
60  READ N
70  FOR I=1 TO N
80  READ D(I),F(I)
90  F=F+F(I)
100 NEXT I
110 FOR I=1 TO N
120 P=0
130 FOR J=1 TO N
140 IF D(I)>D(J) THEN 170
150 N1=D(I)\N2=0
160 GOTO 180
170 N1=D(J)\N2=D(I)-D(J)
180 Z=N1*S1+N2*S2-D(I)*C1
190 P=P+Z*F(J)
200 NEXT J
210 P=P/(100*F)
220 PRINT D(I),P
230 IF P<M THEN 250
240 M=P\L=D(I)
250 NEXT I
260 PRINT "MAXIMUM PROFIT IS $";M; "PER DAY"
270 PRINT "WHEN";L; "LOAVES ARE PRODUCED"
280 DATA 7,70,3,80,11,90,22,100,37
290 DATA 110,19,120,6,130,2
300 END
```

For the original problem the inputs will be C1=13, S1=32, S2=6. On line 50, F, the sum of the frequencies, and M, the maximum expected profit, are initialized (−1E12, using E format, represents minus one billion). The READ statement on the next line will result in the allocation N=7, the number of demands in the table. The single loop from lines 70 to 100 will read in the demands D(I) and the frequencies F(I) from the data list to give the allocations D(1)=70, F(1)=3, D(2)=80, F(2)=11, etc., the sum of the frequencies being evaluated on line 90. The major part of the program between lines 110 and 250 will be illustrated when I=4, and D(I)=D(4)=100. Line 120 sets P, the expected profit, to zero.

J	D(I)	D(J)	N1	N2		Z	F(J)	P
1	100	70	70	30	(line 170)	1120	3	3360
2	100	80	80	20	(line 170)	1380	11	18540
3	100	90	90	10	(line 170)	1640	22	54620
4	100	100	100	0	(line 150)	1900	37	124920
5	100	110	100	0	(line 150)	1900	19	161020
6	100	120	100	0	(line 150)	1900	6	172420
7	100	130	100	0	(line 150)	1900	2	176220

D(I) is the number of loaves made per day, D(J) the possible future demands, N1 the number of loaves sold and N2 the number left at the end of the day. The IF-THEN statement on line 140 ensures that the values of N1 and N2 are correctly evaluated in each of the two possible situations. Z gives the profit made – on the first line of the table it is the profit when 100 loaves are made and the demand is 70. The values of Z are multiplied by the frequency of the demand and added onto P, the sum of all the profits when 100 loaves are made. Line 210 converts P into the average profit in dollars–thus P=176220/10000=$17.622 since F=100. Finally lines 230, 240 are used to save the maximum profit M and the number of loaves L made in this case. For the next value of I, the program will repeat the above procedure for D(5)=110 loaves made (after setting P to zero again on line 120). The program produces the following:

```
COST OF MAKING PRODUCT ? 13
NORMAL SELLING PRICE ? 32
LOWER SELLING PRICE ? 6
70          13.30
80          15.122
90          16.658
100         17.622
110         17.624
120         17.132
130         16.484
MAXIMUM PROFIT IS $ 17.624 PER DAY
WHEN 110 LOAVES ARE PRODUCED.
```

Of course the maximum profit may also be determined using the concepts of marginal probability, marginal loss and cumulative probability.

CORRELATION AND REGRESSION

The past sales figures of a company are as follows:

Year X	Sales Y (thousands)
1	15
2	16
3	18
4	22
5	23
6	25
7	27

To calculate the correlation coefficient R and the regression line of Y on X requires the formulae:

$$R = \frac{S_{XY}}{S_X S_Y} \quad \text{where}$$

the covariance $S_{XY} = \frac{1}{N}\sum(X - \overline{X})(Y - \overline{Y})$

the standard deviation of X is $S_X = \sqrt{\frac{1}{N}\sum(X-\overline{X})^2}$

the arithmetic mean of X is $\overline{X} = \frac{1}{N}\sum X$

and N is the number of X values.
(S_Y and $\overline{Y}$, the standard deviation and mean of Y respectively, are calculated similarly.)
The regression line of Y on X is Y=MX+C where

$$M = \frac{S_{XY}}{S_X^2} \quad \text{and} \quad C = \overline{Y} - M\overline{X}$$

The formula for the standard deviation quoted here tends to give more accurate answers than the one on page 23, even though they are mathematically equivalent. All computers accumulate small errors known as **rounding** or **truncation errors** and these are considerably multiplied when squaring large numbers – the formula above contains the term $(X-\overline{X})^2$ which tends to be smaller than the corresponding term of X^2 in the original formula. Using this formula we have to calculate the arithmetic mean before being able to find the standard deviation, and subscripted variables will be needed to avoid inputting the data twice.

Additional variables

X(I)	the X values	Y(I)	the Y values
M1	the mean of X	M2	the mean of Y
S1	$\sum(X - \overline{X})^2$	S2	$\sum(Y - \overline{Y})^2$
S3	$\sum(X - \overline{X})(Y - \overline{Y})$		

```
10 DIM X(100), Y(100)
15 M1=0\M2=0\S1=0\S2=0\S3=0
20 PRINT "NUMBER OF PAIRED VALUES"
30 INPUT N
40 PRINT "INPUT THE X AND Y VALUES IN PAIRS"
50 FOR I=1 TO N
60 INPUT X(I), Y(I)
70 M1=M1+X(I)
80 M2=M2+Y(I)
90 NEXT I
100 M1=M1/N
110 M2=M2/N
120 FOR I=1 TO N
130 S1=S1+(X(I)-M1)↑2
140 S2=S2+(Y(I)-M2)↑2
150 S3=S3+(X(I)-M1)*(Y(I)-M2)
```

```
160 NEXT I
170 R=S3/SQR(S1*S2)
180 M=S3/S1
190 C=M2-M*M1
200 PRINT "CORRELATION COEFFICIENT IS";R
210 PRINT "REGRESSION LINE IS Y=";M;"*X+";C
220 PRINT TAB(1);"X";TAB(8);"OBSERVED Y";
225 PRINT TAB(20);"PREDICTED Y"
230 FOR I=1 TO N
240 P=M*X(I)+C
250 PRINT TAB(1);X(I);TAB(12);Y(I);TAB(24);P
260 NEXT I
270 PRINT "INPUT ANY X-VALUES FOR Y PREDICTION"
280 PRINT "INPUT -1 TO END"
290 INPUT X;
300 IF X=-1 THEN 340
310 P=M*X+C
320 PRINT "PREDICTION IS";P
330 GOTO 290
340 END
```

Sample run

```
NUMBER OF PAIRED VALUES
? 7
INPUT THE X AND Y VALUES IN PAIRS
? 1,15
? 2,16
? 3,18
? 4,22
? 5,23
? 6,25
? 7,27
CORRELATION COEFFICIENT IS 0.989955
REGRESSION LINE IS Y = 2.10714 *X + 12.4286
X    OBSERVED Y  PREDICTED Y
1         15          14.5357
2         16          16.6428
3         18          18.75
4         22          20.8571
5         23          22.9643
6         25          25.0714
7         27          27.1786
INPUT ANY X-VALUES FOR Y PREDICTION
INPUT -1 TO END
? 8 PREDICTION IS 29.2875
? 9 PREDICTION IS 31.3928
? 10 PREDICTION IS 33.5
? 11 PREDICTION IS 35.6071
? 12 PREDICTION IS 37.7143
? -1
```

The values of S1, S2 and S3 have not been divided by N because N would cancel anyway in the evaluation of R, M and C. The loop from lines 230 to 260 prints the observed and predicted Y values using the regression line in order to inspect the goodness of fit. Further X values can be input to obtain predictions by extrapolation.

SEARCHING

The problem in any search procedure is to locate a particular record in a given data list by an efficient algorithm. For example it may be necessary to try to find if a particular value or name is present among a list of subscripted variables. The simplest technique is known as a **linear search** and is often used when the data in the list is unordered. Suppose that the variables A(I) have already been allocated values within a program as follows: A(1)=6, A(2)=3, A(3)=1, A(4)=7, A(5)=2, A(6)=0, A(7)=9, A(8)=11, A(9)=14. The following procedure will now perform a search for any required value.

```
200 PRINT "SEARCH VALUE";\INPUT X
210 FOR I=1 TO 9
220 IF A(I)<>X THEN 240
230 J=I\GOTO 270
240 NEXT I
250 PRINT "VALUE NOT FOUND"
260 STOP
270 PRINT "VALUE FOUND IN POSITION";J
280 STOP
```

The loop between lines 210 and 240 runs through the list of variables and line 220 checks to see if the **search value** X is equal to A(I), the Ith number in the list. If they are unequal, transfer is made to line 240 to check the next number in the list. In the case of the search value being absent from the list, the program will never reach line 230 and so eventually reaches line 250 on the completion of the loop. As soon as the X value is found in the list, line 230 is executed and J is made equal to I, the position of the search value in the list. Exit is then made from the loop to print the fact that the value has been found.

Dry run when X=2

I	A(I)	A(I) = X?	Transfer to line	J
1	6	No	240	0
2	3	No	240	0
3	1	No	240	0
4	7	No	240	0
5	2	Yes	230	5

If the elements in the list are known to be in numerical order, the procedure can be improved by replacing line 220 by the two lines:

```
220 IF X>A(I) THEN 240
225 IF X<A(I) THEN 250
```

The latter line represents the fact that if the search value is already less than the data value, it must also be less than all later data values and therefore cannot be in the list. The loop is then exited immediately to avoid wasting time in checking the remainder of the list. The average number of comparisons required for an unordered list is (n+1)/2 if the search value is present and n if the search value is absent, where n is the number of values in the list. For the ordered list above the average is (n+1)/2 whether the search value is in the list or not. In practice the list may contain thousands of values and in this case a linear search can be a slow process.

BINARY SEARCH

This technique is much faster than the linear search and reduces loop times considerably. It is used on data lists which are in numerical or alphabetical order. The aim is roughly to halve the number of values to be searched at each comparison. The first check is made in the middle of the list, thus determining that the search value is either in the lower half or the upper half and then the next check is made in the middle of the reduced list and so on.

Dry run to search for value of X=9 in the list A(1)=2, A(2)=3, A(3)=5, A(4)=8, A(5)=9, A(6)=11, A(7)=16.

L	U	I	A(I)	A(I) = X?	A(I) < X?
1	7	4	8	No	Yes
5	7	6	11	No	No
5	5	5	9	Yes	–

Initially the position of the search value (if present) is between 1 and 7 inclusive – L represents the lower position and U the upper. I is the middle position of the interval and is simply the average of L and U. If A(I)=X then the value has been found immediately and the algorithm ends. If A(I)<X the search value X must be in the upper half of the list and so L is changed to 5 as this is now the lowest possible position for X. The procedure is then repeated and in this case A(I)>X which indicates that X is in the lower half of the new interval and U changes to 5 before the next comparison is made.

Two complications can arise in the algorithm. Suppose that the value of 8.5 is being searched for instead of 9. The dry run will proceed exactly as shown in the first two rows but the answers to the two questions in the third row will both be no, resulting in a fourth row of L=5, U=4. This must be avoided as L must obviously always be less than or equal to U. Thus a check to see whether L is larger than U must be inserted at each step, the truth of this statement indicating that the value is absent from the list.

The second complication arises when I is not a whole number. Consider the list A(1)=2, A(2)=5, A(3)=6, A(4)=10, A(5)=13, A(6)=14, A(7)=18, A(8)=20 with a search value of 20.

L	U	I	A(I)	A(I) = X?	A(I) < X?
1	8	4	10	No	Yes
5	8	6	14	No	Yes
7	8	7	18	No	Yes
8	8	8	20	Yes	–

Stop – value found in position 8.

L	U	I	A(I)	A(I) = X?	A(I) < X?
1	8	5	13	No	Yes
6	8	7	18	No	Yes
8	8	8	20	Yes	–

The first example has calculated I by finding the average of L and U and then rounded it down to the nearest whole number if necessary. The second example has rounded up to the next whole number instead. The following **subprogram** will deal with both of these complications.

```
200 PRINT "SEARCH VALUE";\INPUT X
210 L=1\U=N
220 IF L>U THEN 300
230 I=INT((L+U)/2)
240 IF A(I)=X THEN 280
250 IF A(I)<X THEN 270
260 U=I-1\GOTO 220
270 L=I+1\GOTO 220
280 PRINT "VALUE IN POSITION";I
290 STOP
300 PRINT "VALUE NOT FOUND"
310 STOP
```

Here N is the number of variables in the list. It is to be assumed that N and the subscripted variables A(I) have been assigned values before line 200 is executed. Line 220 is the check for the absence of the search value. The middle of the interval is evaluated on the next line using the **integer function** which will be looked at in more detail in the next chapter. It will round down the value of (L+U)/2 when necessary, e.g. INT(4.5)=4 and INT(7)=7.

The binary search will need at most seven comparisons for a list containing 128 values (2 to the power of 7 is equal to 128) compared with an average of approximately 64 for a linear search. In general a list of *n* values requires, at most, *m* comparisons where *m* is the power to which 2 has to be raised to produce the first number greater than or equal to *n* (in fact *m* is proportional to log *n* where the base of the logarithm is 2). For example, a list of 47 data values needs at most six comparisons because 2 to the power of 6 is equal to 64>47 whereas 2 to the power of 5 is equal to 32<47. The binary search procedure is by far the most efficient of the two methods considered here, especially when applied to large lists, though it is not applicable to unordered data. Often the list to be searched is already in order or has been sorted by the algorithm introduced earlier in this chapter. Note that a minor alteration to the last program will deal with data in decreasing order rather than increasing order. Many other **search algorithms** are available but are beyond the scope of this book.

Chapter 7

Functions

Series of special functions are usually supplied with computer software. Examples are SQR(X) and TAB(X) which have already been mentioned. They are really mini-programs because the instruction A = SQR(2) is performed by a special square root program within the computer itself.

MATHEMATICAL FUNCTIONS

The exponential function is of the form EXP(X) and returns the value of e to the power of X, e.g.
EXP(1)=e=2.71828, EXP(0)=1

The logarithmic function is of the form LOG(X) and returns the value of the logarithm of the number X to base e, e.g. LOG(1)=0, LOG(2.71828)=1

To calculate logarithms to the standard base of 10 one can use the formula A = LOG(X)/LOG(10),
e.g. when X = 10, A = 1; when X = 100, A = 2

In a similar way, logarithms to base 2 will be given by the formula A=LOG(X)/LOG(2) and so on for any other base. Any attempt to calculate the logarithm of a non-positive number will lead to an error.

The square root function has the form SQR(X), e.g.
SQR(49)=7, SQR(441)=21

The sine function has the form SIN(X) and returns the sine of an angle X which has to be measured in radians. If the angle X is measured in degrees the above expression has to be replaced by SIN(X*3.14159/180), e.g.
SIN(30*3.14159/180)=0.5 and means sin 30°.
SIN(450*3.14159/180)=1 and means sin 450°.

The cosine function has the form COS(X) and works in the same manner as SIN.

The tangent function has the form TAN(X) when it is available on a system – it can also be calculated by the formula SIN(X)/COS(X).

The arctangent function has the form ATN(X) and returns the angle (in radians) whose tangent is X. To convert this angle to degrees use 180*ATN(X)/3.14159.

SPECIAL FUNCTIONS

The absolute value function ABS(X) gives the absolute value (sometimes called the modulus) of the number X, i.e. the number is made positive, e.g.
ABS(4)=4, ABS(−5)=5, ABS(0)=0.

The sign function SGN(X) returns the value of 1 if X is

positive, 0 if X is zero and −1 if X is negative, e.g. SGN(7)=1, SGN(0)=0, SGN(−9.7)=−1.

These two functions are related to one another by the following formula:
ABS(X) = X*SGN(X)
For example:

X	SGN(X)	X*SGN(X)	ABS(X)
5	1	5	5
0	0	0	0
−2	−1	2	2

The integer function INT(X) returns the largest integer (whole number) which is less than or equal to the number X, e.g. INT(4.3)=4, INT(6)=6, INT(5.91)=5, INT(−3)=−3, INT(−6.2)=−7
This function is needed to round off a data value to the nearest whole number using the formula INT(X+0.5).

X	X + 0.5	INT(X + 0.5)
7.4	7.9	7
7.5	8	8
7.6	8.1	8
4	4.5	4
−3.6	−3.1	−4
−3.5	−3	−3
−3.4	−2.9	−3
−4	−3.5	−4

Note that numbers such as 7.5 and −3.5 are always rounded up.

Expressing a number to two decimal places can be achieved by the formula INT(100*X+0.5)/100.

X	100*X	100*X+0.5	INT	INT/100
7.3626	736.26	736.76	736	7.36
9.4269	942.69	943.19	943	9.43
−3.5214	−352.14	−351.64	−352	−3.52
−6.1991	−619.91	−619.41	−620	−6.2

In general INT(10↑N*X+0.5)/10↑N approximates the number X to N decimal places.

The median of a set of data can be evaluated using the INT function. If the numbers are stored in the variables A(1), A(2), . . . , A(N) and have been placed into order by the sorting algorithm of the last chapter, the following additional lines will print the median.

```
191 IF N/2=INT(N/2) THEN 195
192 K=(N+1)/2
193 M=A(K)
194 GOTO 198
195 K=N/2
196 L=K+1
197 M=(A(K)+A(L))/2
198 PRINT "MEDIAN IS";M
```

For example, if N=6 and A(1)=1, A(2)=3, A(3)=4, A(4)=7, A(5)=7, A(6)=9, the median is the average of the two middle numbers A(3), A(4), i.e. median= (4+7)/2=5.5. Line 191 checks to see whether N is an even or an odd number. In this case we get:

N/2	INT(N/2)	Transfer to line	K	L	M
3	3	195	3	4	5.5

When N=5 and A(1)=2, A(2)=3, A(3)=7, A(4)=8, A(5)=9, the median is the middle number, i.e. A(3)=7.

N/2	INT(N/2)	Transfer to line	K	M
2.5	2	192	3	A(3)=7

Thus lines 192 to 194 find the median when N is odd and lines 195 to 197 when N is even. Note that it is legal to replace lines 192 and 193 by the equation M=A((N+1)/2).

HISTOGRAMS

The next program will take in a set of numbers, calculate the group frequencies and plot the results in the form of a histogram. The data might refer to the sales volume of a particular product on consecutive days.

11	21	17	22	13	23	36	37	19	20	31	14	10	19	22
21	31	32	38	42	24	35	37	25	23	41	12	25	30	40
14	16	17	13	21	11	26	21	15	31	26	29	25	29	28
19	34	26	29	34	45	28	16	31	29	17	28	20	30	39

The user types in the groups to be tabulated – A is the lower endpoint of the first group, B is the upper endpoint of the last group and G is the group interval. If A=10, B=45, G=5 the groups will be 10–15, 15–20, 20–25, 25–30, 30–35, 35–40, 40–45. To code these groups for the program, they will be numbered from 1 to 7. Therefore:
Group 1 is 10–15, i.e. 10+5*0 to 10+5*1
Group 2 is 15–20, i.e. 10+5*1 to 10+5*2
Group 3 is 20–25, i.e. 10+5*2 to 10+5*3
Group 4 is 25–30, i.e. 10+5*3 to 10+5*4, etc.
Group I is therefore 10+5*(I−1) to 10+5*I
In general, for group I, the lower endpoint L=A+G*(I−1) and the upper endpoint U=A+G*I. The program will in fact calculate L by subtracting G from U.

The next problem is to determine the group to which a number is to be assigned.

NumberX	X−10	(X−10)/5	INT((X−10)/5)	Add 1
11	1	0.2	0	1
20	10	2	2	3
35	25	5	5	6
42	32	6.4	6	7

The final column gives the number of the group. In general it has the formula I=INT((X−A)/G)+1. The number 20 has been placed in the group 20–25 and not 15–20 so the convention here is that the group 20–25 means greater than or equal to 20 but less than 25. N is the total number of groups which is (45−10)/5, i.e. N=(B−A)/G.

```
10 DIM F(20)
20 PRINT "FIRST GROUP'S LOWER ENDPOINT"\INPUT A
30 PRINT "FINAL GROUP'S UPPER ENDPOINT"\INPUT B
40 PRINT "GROUP INTERVAL"\INPUT G
50 N=(B-A)/G
60 IF N<>INT(N) THEN 20
70 IF N<=0 THEN 20
80 FOR I=1 TO N
90 F(I)=0
100 NEXT I
110 PRINT "INPUT THE DATA-TYPE -1 TO END LIST"
120 INPUT X
130 IF X=-1 THEN 190
140 IF X<A THEN 120
150 IF X>=B THEN 120
160 I=INT((X-A)/G)+1
170 F(I)=F(I)+1
180 GOTO 120
190 FOR I=1 TO N
200 U=A+G*I\L=U-G
210 PRINT TAB(1);L;TAB(4);"TO";TAB(7);U;TAB(13);F(I);
220 IF F(I)=0 THEN 260
230 FOR J=1 TO F(I)
240 PRINT TAB(J+18);"O";
250 NEXT J
260 PRINT
270 NEXT I
280 END
```

The frequency for group I is F(I) and the first line will permit a maximum of 20 groups. Line 60 checks whether N is a whole number – when N=3, INT(N)=3 and they are equal but when N=4.6, INT(N)=4 and they differ. In the latter case control returns to line 20 to ask for three consistent inputs. Similarly line 70 ensures that only positive values for N can be allowed. Checks are also made on lines 140 and 150 to ignore a data value which lies outside the range input at the beginning. When the group is determined, the relevant frequency is incremented by 1 before returning for the next value. For instance, if I = 4, line 170 becomes F(4)=F(4)+1 meaning add 1 onto the fourth group's frequency. After the input of the rogue value the I loop will generate the rows of the graph, one row for each group. The J loop prints F(I) 0's to represent the frequency for the I'th group. When J=1 the TAB position is 19, when J=2 it is 20, etc. Line 220 avoids this loop if the frequency is zero. Only the histogram for the sample run is shown.

```
10 TO 15   8   00000000
15 TO 20   9   000000000
20 TO 25   11  00000000000
25 TO 30   13  0000000000000
30 TO 35   9   000000000
35 TO 40   6   000000
40 TO 45   4   0000
```

RANDOM NUMBERS

A table of random numbers or a pseudo-random number generator can be called. The function RND(T) will produce a random number in the range 0 to 0.999999 from the computer's list of numbers (or from a special sequence designed to give near random numbers). The subscript T has no significance here and RND(X) or **RND(Ø)** would equally suffice – sometimes it is permissible to drop the subscript altogether and just use RND. The expression RND(Ø) will be used in all programs detailed in the remainder of the book.

```
1Ø  RANDOMIZE          Output   .569325
2Ø  FOR I=1 TO 3                .684777
3Ø  A=RND(Ø)                    .139982
4Ø  PRINT A
5Ø  NEXT I
6Ø  END
```

The above program produces the three random numbers shown. The **RANDOMIZE** statement on line 1Ø has the effect of producing a different set of random numbers when the program is run again. It has to be placed before the occurrence of RND on line 3Ø and is probably best written at the beginning of the program or just after any DIM statements. By omitting line 1Ø the program would produce the same set of random numbers on each consecutive run – this would be useful if the program had an error in it and the programmer wanted to correct a few lines and rerun with the identical data. It is possible to combine the INT and the RND functions to produce random sets of integers over any range.

RND(Ø) gives a random number between 0 and 0.999999.
6*RND(Ø) gives a random number between 0 and 5.99999.
INT(6*RND(Ø)) gives a random integer between 0 and 5.
INT(6*RND(Ø))+1 gives a random integer between 1 and 6 and could be used to simulate the throw of a dice.

Similarly:
INT(52*RND(Ø)) produces a random integer from 0 to 51
INT(1Ø*RND(Ø))+1 produces a random integer from 1 to 10.
INT (16*RND(Ø))+12 produces a random integer from 12 to 27.

QUEUING SIMULATION

The random number generator may be used to simulate inter-arrival and service times in the modeling of queues. At this stage only the simplest case of a single server will be considered. Suppose that the problem is to investigate the arrival of trucks at a single loading bay and that the following table is the past record of 100 trucks.

Inter-arrival time (minutes)	Frequency	Loading time (minutes)	Frequency
3	9	6	18
5	11	7	36
7	40	8	32
9	31	9	14
11	9		

The variables used in the program are:

A(I)	inter-arrival times	F(I)	arrival frequencies
B(I)	the loading times	G(I)	loading frequencies
Z1	no. of arrival times	Z2	no. of loading times
M	length of simulation	T	the clock time
F	flag variable	A	time of next arrival
R	a random number	N	number of arrivals
S	no. of departures	B	next departure time
Q	queue length (including the bay)		
W	total waiting time (incl. loading time)		
I1	total idle time		

```
10 PRINT "QUEUING SIMULATION"
20 DIM A(10),B(10),F(10),G(10)
30 RANDOMIZE
40 Z1=5\Z2=4
50 F(0)=0\G(0)=0
60 FOR I=1 TO Z1\READ A(I),F(I)
70 F(I)=F(I)+F(I-1)\NEXT I
80 FOR I=1 TO Z2\READ B(I),G(I)
90 G(I)=G(I)+G(I-1)\NEXT I
100 PRINT "LENGTH OF SIMULATION IN MINUTES"\INPUT M
110 A=0\Q=0\N=0
120 B=-1\S=0\I1=0\W=0
130 PRINT TAB(19);"CLOCK TIME";TAB(32);"QUEUE LENGTH"
140 FOR T=0 TO M
150 F=0\W=W+Q
160 IF A>T THEN 250
170 Q=Q+1\N=N+1
180 PRINT "ARRIVAL";\F=1
190 R=F(Z1)*RND(0)
200 FOR I=1 TO Z1
210 IF R>=F(I) THEN 230
220 A=A(I)\GOTO 240
230 NEXT I
240 A=A+T
250 IF B>T THEN 360\IF B<T THEN 280
260 Q=Q-1\S=S+1
270 PRINT TAB(9);"DEPARTURE";\F=1
280 IF Q>0 THEN 300
290 I1=I1+1\GOTO 360
```

```
3ØØ  R=G(Z2)*RND(Ø)
31Ø  FOR I=1 TO Z2
32Ø  IF R>=G(I) THEN 34Ø
33Ø  B=B(I)\GOTO 35Ø
34Ø  NEXT I
35Ø  B=B+T
36Ø  IF F=Ø THEN 38Ø
37Ø  PRINT TAB(24);T;TAB(37);Q
38Ø  NEXT T
39Ø  PRINT\PRINT "NUMBER OF ARRIVALS";N
4ØØ  PRINT "NUMBER OF DEPARTURES";S
41Ø  PRINT "NUMBER STILL IN QUEUE";Q
42Ø  PRINT "TOTAL WAITING TIME";W;"AVERAGE";W/S;"MINS"
43Ø  PRINT "TOTAL TIME BAY IDLE";I1;"MINS"
44Ø  DATA 3,9,5,11,7,4Ø,9,31,11,9
45Ø  DATA 6,18,7,36,8,32,9,14
46Ø  END
```

Z1 and Z2 are given their initial values on line 4Ø and are written as Z1,Z2 and not 5,4 in the rest of the program. Only the three lines 4Ø,44Ø,45Ø need to be altered to rerun the program with a different set of sampling times. The first task is to read the values from the data list and place them in the appropriate subscripted variables. The statement of line 7Ø converts the frequencies F(I) into cumulative frequencies in the following manner.

I	**A(I)**	**F(I)**	**Line 7Ø**	**F(I)**
1	3	9	F(1) = F(1) + F(0) = 9 + 0	9
2	5	11	F(2) = F(2) + F(1) = 11 + 9	20
3	7	40	F(3) = F(3) + F(2) = 40 + 20	60
4	9	31	F(4) = F(4) + F(3) = 31 + 60	91
5	11	9	F(5) = F(5) + F(4) = 9 + 91	100

The variables F(I) now contain the cumulative frequencies, the actual frequencies having been lost (they can be recalculated from the cumulative frequencies if required). Line 5Ø ensures that F(I) retains its initial value on the first pass through the loop. F(0) is used solely for this purpose and takes no further part in any calculations. The sum of the frequencies (100) is now stored in F(5), i.e. F(Z1). The sampling of an inter-arrival time is performed by the loop from lines 2ØØ to 23Ø using the linear search method of the last chapter. The random number RND(Ø) is first multiplied by F(Z1)=F(5)=100 so that R (line 19Ø) is a number chosen at random in the range 0 to 99.9999. The loop then searches through the cumulative frequencies F(I) to find the first value of F(I) which is greater than R.

Dry run when R=100*0.735619=73.5619

I	**F(I)**	**R≥F(I)?**	**Transfer to line**	**A**
1	9	Yes	23Ø	–
2	20	Yes	23Ø	–
3	60	Yes	23Ø	–
4	91	No	22Ø	A=A(I)=A(4)=9

Here the value A=9 minutes has been sampled as the next inter-arrival time. The actual clock time T is then added onto A to give the time of the next arrival (line 240). This procedure will sample according to the given frequencies as follows:

A = 7 is sampled when $20 \leq R < 60$, i.e. probability 0.40
A = 9 is sampled when $60 \leq R < 91$, i.e. probability 0.31
A = 11 is sampled when $91 \leq R < 100$, i.e. probability 0.09
A = 9 is sampled when $60 \leq R < 91$, i.e. probability 0.31
A = 11 is sampled when $91 \leq R < 100$, i.e. probability 0.09

The sampling of the loading times is performed in exactly the same fashion between lines 300 and 350. The remainder of the variables are initialized on lines 110 and 120. The next arrival time A is set to zero and guarantees that the simulation begins with an arrival at time zero. The next departure time (i.e. the time when the loading bay becomes vacated) is set to −1 to ensure that a loading time is sampled on the first pass through the main loop between lines 140 and 380. For each value of the clock time T between 0 and M, W is increased by the queue length to calculate the waiting time so far, and then checks are made for the following conditions:

A>T Next arrival time after present time T, check departures on line 250.
A=T Next truck arrives, increase Q and N by 1 (line 170) and sample next arrival time before checking departures.
B>T Next departure time after present time T, go to line 360 to print results.
B=T Truck fully loaded and departs, decrease Q by 1 and increase S by 1 (line 260), then check queue length (line 280).
B<T Check queue length. This condition occurs when the bay has been idle, i.e. there is no next departure time and B still contains the last departure time which has already happened.
Q>0 Sample a new loading time on line 300 – this occurs when there has been a departure and there is still a truck in the queue or when there has been an arrival to an empty queue.
Q=0 Queue empty, increase I1 and avoid sampling a new loading time (i.e. either no more arrivals and the queue is empty or the last departure was the only truck in the queue).

On every pass through the loop, line 360 is eventually reached. The variable F is called a **flag variable** or **switch variable**. Its function in this program is to print the results if it is set and to avoid printing them if it is unset. It is assigned the value zero by line 150 on every pass through the main loop and can only be set to one when an arrival or departure occurs on lines 180 or 270. Line 360 then checks the flag and the results are not printed when an arrival or a departure has not occurred (F=0).

Sample run

```
QUEUING SIMULATION
LENGTH OF SIMULATION IN MINUTES
? 60
                              CLOCK TIME   QUEUE LENGTH
ARRIVAL                           0              1
ARRIVAL                           7              2
            DEPARTURE             8              1
ARRIVAL                          14              2
            DEPARTURE            15              1
            DEPARTURE            22              0
ARRIVAL                          25              1
ARRIVAL     DEPARTURE            32              1
ARRIVAL                          37              2
            DEPARTURE            41              1
ARRIVAL                          46              2
            DEPARTURE            49              1
ARRIVAL                          51              2
            DEPARTURE            56              1
ARRIVAL                          58              2

NUMBER OF ARRIVALS 9
NUMBER OF DEPARTURES 7
NUMBER STILL IN QUEUE 2
TOTAL WAITING TIME 73 AVERAGE 10.4285 MINS
TOTAL TIME BAY IDLE 3 MINS
```

The calculation of the average time as total waiting time divided by the number of departures is an approximation because two trucks are still in the queue at the end of the run. This average includes the loading time which has an average value of about 7.5 minutes, leaving a time of three minutes between a truck arriving and going into the loading bay. In a simulation of 10000 minutes, which took the computer about four minutes to run, the maximum queue length was six, four trucks were waiting at the end, the average waiting time was 25 minutes and the bay was idle for 148 minutes. Any decision concerning the possible addition of another loading bay would have to take account of these figures.

USER DEFINED FUNCTIONS

In addition to the standard functions outlined in this chapter, the programmer can define his own by using the **define statement DEF FNA(X)**=X↑2+X+1.
DEF is an abbreviation for define; FNA is the name of the function and must begin with the letters FN followed by an alphabetic letter. Up to 26 functions may be defined in a program as FNA, FNB, FNC, . . . , FNZ. X is a dummy variable which indicates that the function is a formula in terms of the letter X. In this example the function FNA

calculates the sum of the square of X, X and 1. Whenever FNA occurs later in the program it will be evaluated according to the definition. For example:

```
10 DEF FNA(X)=X↑2+X+1
20 PRINT FNA(3)
30 Z=4
40 C=FNA(Z)
50 PRINT C
60 END
```

Output

```
13
21
```

Here FNA(3)=3↑2+3+1=9+3+1=13 and
C=FNA(Z)=FNA(4)=4↑2+4+1=16+4+1=21

It is obviously necessary to define a function before using it in a program. Other examples of user defined functions are:
DEF FNB(T)=INT(100*T+0.5)/100 which will round the numbers T to two decimal places – here the dummy variable is T. This would be a concise method of expressing all output to this accuracy – a print statement such as PRINT A would simply be replaced by PRINT FNB(A).
DEF FNC(T)=INT(10↑N*T+0.5)/10↑N will round to N decimal places. The variable N would have been input or assigned a value before this define statement.
DEF FND(A)=LOG(A)/LOG(10) will give the logarithm of A in base 10.
Defining a function in terms of a previously defined user function is also permitted:
DEF FNE(X)=FNA(X)*X
DEF FNF(X)=FNB(FND(X)) meaning find the logarithm of X to base 10 and express to two decimal places.
The function may have more than one variable:
DEF FNG(X,Y)=(X+Y)/2 will find the average of the two numbers X and Y.
DEF FNH(A,B,C)=(A*B*C)↑1/3 finds the cube root of the product of the three numbers, i.e. the geometric mean.
Thus FNG(4,12)=8 and FNH(2,9,12)=6.
The user functions cannot be inputs to a program and changing a function requires that the relevant line be retyped.

GRAPH PLOTTING

The user defined function can be used to plot the graph of any function which can be expressed as a formula in a single variable. The program asks for the range of values to be plotted, calculates the coordinates, decides upon a suitable scale and plots axes if necessary.

```
10 PRINT "GRAPH PLOTTING"
20 PRINT "TO ALTER THE FUNCTION,RE-TYPE LINE 40"
30 DIM Y(100)
40 DEF FNA(X)=0.398942*EXP(-0.5*X*X)
50 PRINT "START,FINISH,STEP FOR X VALUES"
60 INPUT A,B,S
70 C=0\M1=1E99\M2=-M1\Z=0
80 FOR X=A TO B STEP S
```

```
 90  C=C+1\Y(C)=FNA(X)
100  IF Y(C)>M1 THEN 120
110  M1=Y(C)
120  IF Y(C)<M2 THEN 140
130  M2=Y(C)
140  NEXT X
150  PRINT "MINIMUM VALUE";M1;"MAXIMUM VALUE";M2
160  PRINT "RANGE OF Y VALUES TO BE PLOTTED"
170  PRINT "FROM";\INPUT Y1;\PRINT "TO";\INPUT Y2
180  W=Y2-Y1
190  IF SGN(Y1*Y2)>0 THEN 220
200  Z=7-50*Y1/W
210  Z=INT(Z+0.5)
220  PRINT TAB(7);Y1;TAB(57);Y2
230  C=0\FOR X=A TO B STEP S
240  PRINT TAB(1);X;\C=C+1
250  IF Y(C)<Y1 THEN 380\IF Y(C)>Y2 THEN 380
260  T=50*(Y(C)-Y1)/W+7
270  T=INT(T+0.5)
280  IF X=0 THEN 330\IF Z=0 THEN 300
290  IF T<Z THEN 310\IF T>Z THEN 320
300  PRINT TAB(T);"*"\GOTO 390
310  PRINT TAB(T);"*";TAB(Z);"I"\GOTO 390
320  PRINT TAB(Z);"I";TAB(T);"*"\GOTO 390
330  IF T=7 THEN 350
340  FOR I=7 TO T-1\PRINT TAB(I);"-";\NEXT I
350  PRINT TAB(T);"*";
360  IF T=57 THEN 380
370  FOR I=T+1 TO 57\PRINT TAB (I);"-";\NEXT I
380  PRINT
390  NEXT X
400  END
```

Line 40 is the define statement for the normal (or Gaussian) distribution in this run. C is a counter to record the number of points, M1 is the minimum value of the function (set very large initially), M2 the maximum and Z a variable to store the TAB position of the vertical axis (if any). For each value of X the counter is increased by one and the value of the function stored in the subscripted variable Y(C). The first value of the function will go into Y(1), the next into Y(2) and so on. The minimum and maximum values are then printed and the user asked to input the range of Y values to be plotted. W is the width of the graph and is needed to determine the TAB position of a point in the formula on line 260. The condition on line 190 is satisfied when Y1 and Y2 are both positive or both negative and in this case Z retains its value of 0, to indicate that no vertical axis is to be printed. Otherwise the TAB position for this axis is evaluated on lines 200 and 210. The graph is plotted between positions 7 and 57, values outside the range input being discounted on line 250. The conditions on lines 280 and 290 check for the various possibilities – when X=0 the horizontal axis has to be printed as well as the point, when Z=0 the vertical axis is not printed; otherwise the position of the point and this axis have to be compared (line 290). A sample run follows:

```
GRAPH PLOTTING
TO ALTER THE FUNCTIONS, RE-TYPE LINE 40
START, FINISH, STEP FOR X VALUES
?-2,2,0.2
MIMIMUM VALUE 0.0539909 MAXIMUM VALUE 0.398942
RANGE OF Y VALUES TO BE PLOTTED
FROM ? 0 TO ? 0.4
```

```
     0                                                  0.4
-2   I  *
-1.8 I      *
-1.6 I          *
-1.4 I                *
-1.2 I                      *
-1   I                             *
-0.8 I                                    *
-0.6 I                                           *
-0.4 I                                                *
-0.2 I                                                   *
0    .....................................................*
0.2  I                                                   *
0.4  I                                                *
0.6  I                                           *
0.8  I                                    *
1    I                             *
1.2  I                      *
1.4  I                *
1.6  I          *
1.8  I      *
2    I  *
```

Chapter 8

Arrays

It is often useful and desirable to store numbers in the form of a table rather than a list. These tables are commonly known as **arrays** or **matrices**. The name of the array is a single alphabetic letter and is followed by the two subscripts as in the variable A(I,J). Thus A(3,7) is the number in the third row and seventh column of the array or table. A(I,J) is called a **double subscripted variable** and represents a generalization of the **single subscripted variable** A(I) covered in an earlier chapter. Arrays may be input using a double loop:

```
10  FOR I=1 TO 4
20  FOR J=1 TO 3
30  READ A(I,J)
40  NEXT J
50  NEXT I
60  DATA 5,3,2,4,1,6
70  DATA 2,4,7,3,3,4
80  END
```

Dry run

I	J	A(I,J)	I	J	A(I,J)
1	1	5	3	1	2
1	2	3	3	2	4
1	3	2	3	3	7
2	1	4	4	1	3
2	2	1	4	2	3
2	3	6	4	3	4

This will result in the table:

A(1,1)=5	A(1,2)=3	A(1,3)=2	or simply	5	3	2
A(2,1)=4	A(2,2)=1	A(2,3)=6		4	1	6
A(3,1)=2	A(3,2)=4	A(3,3)=7		2	4	7
A(4,1)=3	A(4,2)=3	A(4,3)=4		3	3	4

The subscript I refers to the number of the row and J to the number of the column. Note that the assignment would be different if the I and J loops were nested the other way around.

In general an array must be **declared** using a DIM statement though some computer systems do allow a double subscripted variable of up to 10 rows and 10 columns to be used without any **declaration**. For larger tables such as one with 11 rows and 13 columns the statement DIM B(11,13) would be needed. In fact this will allocate memory for an array with a row and column zero so that entries such as B(0,0), B(3,0) and B(0,6) may be assigned values if required. DIM B(11,13) will allow a total of $12 \times 14 = 168$ entries and $11 \times 13 = 143$ entries when row zero and column zero are not being used. This dimension statement does not mean that the array must have 11 rows and 13 columns – in some runs only 5 rows and 8 columns may be allocated values.

TABULATING VARIABLES

Double subscripted variables can be operated upon in exactly the same manner as lists and simple variables as the next illustration shows. A company produces five products whose selling prices are given below.

Product	1	2	3	4	5
Price($)	7.25	10.99	11.00	14.99	8.50

The sales quantities during the four quarters of the year are given by the table:

		Product				
		1	2	3	4	5
Quarter	1	84	53	21	32	47
	2	62	41	19	21	41
	3	73	45	20	22	42
	4	69	50	22	25	42

The program will calculate the sales for each quarter, the gross sales, and the quantity and sales for each of the five products during the whole year.

```
10 DIM Q(10,10),P(10)
20 PRINT "NUMBER OF SALES PERIODS";\INPUT M
30 PRINT "NUMBER OF PRODUCTS";\INPUT N
40 FOR I=1 TO M
50 FOR J=1 TO N
60 PRINT "PERIOD";I;"PRODUCT";J;"ITEMS SOLD";
70 INPUT Q(I,J)
80 NEXT J
90 NEXT I
100 FOR I=1 TO N
110 PRINT "SELLING PRICE OF PRODUCT";I;"IN DOLLARS";
120 INPUT P(I)
130 NEXT I
140 T=0
150 FOR I=1 TO M
160 S=0
170 FOR J=1 TO N
180 S=S+P(J)*Q(I,J)
190 NEXT J
200 PRINT "PERIOD";I;"SALES $";S
210 T=T+S
220 NEXT I
230 PRINT "TOTAL SALES $";T
240 FOR J=1 TO N
250 Q=0
260 FOR I=1 TO M
270 Q=Q+Q(I,J)
280 NEXT I
290 S=Q*P(J)
300 PRINT "PRODUCT";J;"NO. ITEMS";Q;"SALES $";S
310 NEXT J
320 END
```

The table is stored in the array Q(I,J) and the prices in the variables P(I). The declaration of these is made on line 10 and allocates more memory than this particular example requires. During the input of the table it is important to state clearly the meaning of the next entry to be input. Line 60 prints the number of the period and the product before asking for the next sales quantity and avoids all possibility of the table being input in the incorrect order. T represents the total sales and is set to zero before the double loop between lines 150 and 220 is executed. For each value of I, S (the sales for a particular period) is also set to zero before its value is calculated by the J loop between lines 170 and 190. For example, when I=1 the run proceeds as follows:

J	P(J)	Q(I,J)	P(J)*Q(I,J)	S
1	7.25	84	609	609
2	10.99	53	582.47	1191.47
3	11.00	21	231	1422.47
4	14.99	32	479.68	1092.15
5	8.50	47	399.50	2301.65

The calculation of the number of items sold and the sales for each product can only be achieved by a double loop in which the J loop is the outer one because J represents the number of the product. Lines 240 to 310 perform this part of the problem and are based upon the same idea as the previous double loop except that the sales (S) for a particular product can be calculated by multiplying the number of items sold (Q) by the price (P(J)).

Sample run (output only)

```
PERIOD 1 SALES $ 2301.65
PERIOD 2 SALES $ 1772.38
PERIOD 3 SALES $ 1930.58
PERIOD 4 SALES $ 2023.5
TOTAL SALES $ 8028.11
PRODUCT 1 NO. ITEMS 288 SALES $ 2088
PRODUCT 2 NO. ITEMS 189 SALES $ 2077.11
PRODUCT 3 NO. ITEMS 82 SALES $ 902
PRODUCT 4 NO. ITEMS 100 SALES $ 1499
PRODUCT 5 NO. ITEMS 172 SALES $ 1462
```

MATRIX MULTIPLICATION

The number of employees required in the various departments of three types of store (large, medium, small) is given as:

		Large	Medium	Small
Department	1	12	14	6
	2	18	9	3
	3	18	6	10

The number of each type of store to be located in New York and California is given as:

		New York	California
Store	Large	23	12
	Medium	16	13
	Small	17	14

It is required to compile a matrix giving the number of employees in each department in New York and California. For example the number of employees required in department 1 in New York is 12×23+14×16+6×17=602. The remaining values are best calculated by multiplying the matrix corresponding to the first table by that of the second table.

```
10 PRINT "MATRIX MULTIPLICATION"
20 DIM A(15,15),B(15,15),C(15,15)
30 PRINT "NO. OF ROWS AND COLUMNS OF FIRST MATRIX"
40 INPUT L,M
50 PRINT "NO. OF ROWS AND COLUMNS OF SECOND MATRIX"
60 INPUT M1,N
70 IF M=M1 THEN 100
80 PRINT "MATRICES CANNOT BE MULTIPLIED"
90 GOTO 30
100 PRINT "INPUT THE FIRST MATRIX"
110 FOR I=1 TO L\FOR J=1 TO M
120 PRINT "ROW";I;"COLUMN";J;\INPUT A(I,J)
130 NEXT J\NEXT I
140 PRINT "INPUT THE SECOND MATRIX"
150 FOR I=1 TO M\FOR J=1 TO N
160 PRINT "ROW";I;"COLUMN";J;\INPUT B(I,J)
170 NEXT J\NEXT I
180 FOR I=1 TO L\FOR J=1 TO N
190 C(I,J)=0
200 FOR K=1 TO M
210 C(I,J)=C(I,J)+A(I,K)*B(K,J)
220 NEXT K
230 NEXT J\NEXT I
240 PRINT "PRODUCT OF THE MATRICES IS"
250 FOR I=1 TO L
260 FOR J=1 TO N
270 PRINT TAB(7*J-6);C(I,J);
280 NEXT J
290 PRINT
300 NEXT I
310 END
```

Sample run (output only for original problem)

```
PRODUCT OF THE MATRICES IS
602     410
609     375
680     434
```

The two matrices can only be multiplied when the number of columns of the first matrix is equal to the number of rows of the second. Line 70 will only allow the input of the two matrices when this condition is satisfied. The answer is stored in the variable C(I,J) which is calculated in the main loop from line 18Ø to line 23Ø.

Dry run
L=3, M=M1=3, N=2

I	J	K	A(I,K)	B(K,J)	C(I,J)
1	1	1	A(1,1)=12	B(1,1)=23	C(1,1)= 0+12×23=276
1	1	2	A(1,2)=14	B(2,1)=16	C(1,1)= 276+14×16=500
1	1	3	A(1,3)=6	B(3,1)=17	C(1,1)= 500+ 6×17=602
1	2	1	A(1,1)=12	B(1,2)=12	C(1,2)= 0+12×12=144
1	2	2	A(1,2)=14	B(2,2)=13	C(1,2)= 144+14×13=326
1	2	3	A(1,3)=6	B(3,2)=14	C(1,2)= 326+ 6×14=410

etcetera

It is possible to place the printing of the answer (lines 25Ø to 3ØØ) in the same loop as the one above and avoid the use of the subscripted variable C(I,J) altogether which would then be replaced by a simple variable. In this way each entry of the product matrix is printed immediately after its calculation and then set to zero. Any problem involving further analysis of the matrix product would have to use the method illustrated in the program given. Some versions do include a special instruction which will automatically multiply two matrices.

REPLACEMENT TECHNIQUES

A company has 200 light bulbs (or any other component which is prone to failure) in operation and wishes to decide upon the optimal replacement procedure. The cost of replacing an individual bulb is $1.50 and the cost per bulb of a group replacement of all 200 bulbs is 30 cents. The statistical characteristics of the life of these particular bulbs is given by the following table:

Week Number	Number of bulbs still working at end of week
0	100
1	80
2	70
3	50
4	30
5	0

The company has the option of either (a) an individual replacement policy, i.e. replacing bulbs only when they fail or (b) a group replacement policy, i.e. replacing all 200 bulbs at regular intervals (even though some may not have failed) and replacing individually where necessary. Determine the optimal type of policy and the interval between group replacements if applicable.

This problem may be solved using a failure tree analysis and a comparison of the costs of the various options open to the company. The table will be stored in the list B(I) so that, in this example, B(0)=100, B(1)=80, B(2)=70, etc. The mean life of the bulb is approximately the sum of these numbers divided by B(0), i.e. M=330/100=3.3. The expected cost per week of the individual replacement policy is then given by C1=200×1.50/3.3=\$90.9091. In order to generate the failure tree, P(I), the proportion of bulbs I weeks old, surviving another week at least, has to be evaluated:

P(0)=80/100=0.8	P(1)=70/80=0.875
P(2)=50/70=0.714286	P(3)=30/50=0.6
P(4)=0/30=0	P(I)=B(I+1)/B(I) in general

The double subscripted variable A(I,J) is used to store the number of bulbs, J weeks old, which are in operation at the beginning of week I. Therefore:

A(0,0)=200	initial number of bulbs
A(1,1)=200×0.8=160	no. surviving the first week
A(1,0)=200−160=40	no. of failures to be replaced
A(2,2)=160×0.875=140	survivals from original 200
A(2,1)=40×0.8=32	survivals from 40 replacements
A(2,0)=200−(140+32)	no. of failures to be replaced

A careful inspection of the location of the numbers in these examples reveals that, in general
A(I,J)=A(I−1,J−1)*P(J−1) when J is not zero, e.g. A(2,1)=A(1,0)*P(0)=40*0.8=32 when I=2, J=1 and A(I,0) is 200 minus the total number of survivals at the beginning of week I.
Note Variables such as A(2,3) have no meaning in this problem and will not be assigned values in the program.

```
10 DIM B(10),P(10),A(10,10)
20 PRINT "REPLACEMENT ANALYSIS"
30 PRINT "NUMBER OF COMPONENTS IN OPERATION";
40 INPUT N
50 PRINT "REPLACEMENT COST PER BULB"
60 PRINT "$";\INPUT C
70 PRINT "COST OF GROUP REPLACEMENT(PER COMPONENT)"
80 PRINT "$";\INPUT G
90 PRINT "INPUT THE STATISTICAL OBSERVATIONS"
100 W=0\S=0
110 PRINT "NO. COMPONENTS WORKING AT END OF WEEK";W;
120 INPUT B(W)\S=S+B(W)
130 IF B(W)=0 THEN 150
140 W=W+1\GOTO 110
150 M=S/B(0)
160 C1=N*C/M
170 FOR I=0 TO W-1
180 P(I)=B(I+1)/B(I)
190 NEXT I
200 A(0,0)=N
210 FOR I=1 TO W
220 S=0
230 FOR J=1 TO I
```

```
24Ø  A(I,J)=A(I-1,J-1)*P(J-1)
25Ø  S=S+A(I,J)
26Ø  NEXT J
27Ø  A(I,Ø)=N-S
28Ø  PRINT A(I,Ø);"FAILURES DURING WEEK";I
29Ø  NEXT I
3ØØ  PRINT
31Ø  PRINT "INTERVAL";TAB(1Ø);"TOTAL COST";
32Ø  PRINT TAB(22);"AVERAGE COST PER WEEK"
33Ø  PRINT "NONE";TAB(12);C1;TAB(29);C1
34Ø  S=Ø\Z=C1\K=Ø
35Ø  FOR I=1 TO W
36Ø  S=S+A(I,Ø)
37Ø  C1=S*C+G*N\C2=C1/I
38Ø  PRINT TAB(2);I;TAB(12);C1;TAB(29);C2
39Ø  IF C2>=Z THEN 41Ø
4ØØ  Z=C2\K=I
41Ø  NEXT I
42Ø  PRINT\PRINT "OPTIMAL POLICY"
43Ø  IF K=Ø THEN 47Ø
44Ø  PRINT "GROUP REPLACEMENT EVERY";K;"WEEKS"
45Ø  PRINT "REPLACING INDIVIDUALLY WHEN NECESSARY"
46Ø  GOTO 48Ø
47Ø  PRINT "INDIVIDUAL REPLACEMENT ONLY"
48Ø  PRINT "AVERAGE COST PER WEEK $";Z
49Ø  END
```

Sample run (output only)

```
4Ø  FAILURES DURING WEEK 1
28  FAILURES DURING WEEK 2
49.6  FAILURES DURING WEEK 3
6Ø.72  FAILURES DURING WEEK 4
9Ø.7Ø39  FAILURES DURING WEEK 5
```

INTERVAL	TOTAL COST	AVERAGE COST PER WEEK
NONE	9Ø.9Ø91	9Ø.9Ø91
1	12Ø	12Ø
2	162	81
3	236.4	78.8
4	327.48	81.87
5	463.536	92.7Ø71

```
OPTIMAL POLICY
GROUP REPLACEMENT EVERY 3 WEEKS
REPLACING INDIVIDUALLY WHEN NECESSARY
AVERAGE COST PER WEEK $78.80
```

Line 13Ø is used to ascertain the input of the final B(I) value, W, then containing the number of weeks during which all the bulbs have failed. Once the number of failures for each week have been calculated the final loop from line 35Ø to line 41Ø will work out the costs of each policy. S is the total number of failures during the interval I between group replacements, Z the minimum cost (set to the cost of individual replacement initially) and K the corresponding interval (set to zero initially).

Chapter 9

String Variables

In this chapter the storage and analysis of other forms of data will be considered. Three basic types of information may be distinguished:

Numeric Any number such as 3, 43Ø, 32.65, 15E−7.
Alphabetic Words or names as in JANUARY, JOHN SMITH. Here only letters in the range A to Z are permitted.
Alphanumeric Expressions such as the car number VFW 981J, the name J SMITH or the formula (A+B)/2. Here the permissible characters are any digit Ø to 9, any letter A to Z and any punctuation marks available ?, +, :, \, $, etc.

A **string** is any group of alphanumeric characters as in the examples above. These strings can be input to the computer and stored in **string variables** which are recognized as a letter followed by a dollar sign or a letter followed by a single digit and a dollar sign, e.g. A$, C7$, Z2$ are permitted string variables. The following lines might be inserted at the end of a program to allow the possibility of a rerun:

```
500 PRINT "DO YOU WANT A RERUN (TYPE YES OR NO)"
510 INPUT A$
520 IF A$="YES" THEN 10
530 END
```

The input on line 51Ø is a word and must be stored in the string variable A$. The statement on line 52Ø checks to see whether A$ contains the word YES and transfers accordingly. Note the quotation marks on this line – all strings in a program must be enclosed by them, though this does not apply to strings input at the keyboard. DATA statements can now contain both numeric and alphanumeric information as in the next example:

```
10 READ N$,A,D$,S$
20 PRINT N$;"AGE";D.O.B.";D$;" ";S$
30 DATA "J. SMITH",31,"3/29/51"· "MALE"
40 END
```

Output

```
J. SMITH AGE 31 D.O.B. 3/29/51 MALE
```

N$ is the name of the person, A the age in years, D$ the date of birth and S$ the sex. The number 31 in the DATA statement is not enclosed by quotation marks because it is a numerical quantity to be stored in the simple variable A. In fact numbers can be stored as string variables if required but this would exclude the possibility of applying arithmetic operations to them unless they were first converted into numeric variables (see the section on string functions in this chapter). The words AGE and D.O.B. on line 2Ø have not been stored in string variables because they do not vary from person to person and can thus form part of a PRINT statement. A blank space has been inserted between the two string variables D$ and S$ on this line because otherwise the strings would be printed without a space between them. This only applies to string variables when the semicolon is being used for small spacing. TAB statements can be inserted to tidy up the output and a loop could be added (plus more data) to print a list of names. Restrictions on the number of characters in a string variable do vary from computer to computer. It will be assumed that the maximum number of characters permitted is 72 and that any string variable requiring more than 8 characters requires a declaration in a dimension statement, e.g. 10 DIM A$(25),B$(16),C$(44). Here A$ may have up to 25 characters, B$ up to 16 characters and C$ up to 44. A$, B$, C$ are simple string variables (or **string constants**), i.e. single strings even though they could contain a sentence. Note the distinction between this and DIM A(25) which declares a list of variables A(1), A(2), . . . , A(25).

SUBSCRIPTED STRING VARIABLES

String variables can be stored in a list with single subscripts but not as a table with two subscripts. The names of the months could be stored in the list of subscripted string variables M$(1), M$(2), . . . , M$(12) referred to as M$(I) in general, e.g. M$(1)="JANUARY", M$(5)="MAY", etc. A dimension statement is required if the list contains more than 10 strings or some of the strings contain more than 8 characters. For the months it would have the form DIM M$(12,9) reserving memory for up to 13 strings with at most 9 characters in each (13 strings because M$(0) may be used). The user should check his reference manual to make sure double subscripted string variables are handled this way. The next program is an example of an extremely simple **information retrieval system**. It contains the sales figures of a company for each month of the past four years and allows the user to access a particular record.

```
10  DIM M$(12,9),S(12,4),A$(9)
20  PRINT "SALES INFORMATION"
30  READ Y
40  FOR I=1 TO 12
50  READ M$(I)
60  FOR J=1 TO 4
70  READ S(I,J)
80  NEXT J
90  NEXT I
100 PRINT "MONTH";\INPUT A$;
110 PRINT "YEAR";\INPUT Y1
120 FOR I=1 TO 12
130 IF A$<>M$(I) THEN 150
140 I1=I\GOTO 170
150 NEXT I
160 PRINT "MONTH NOT KNOWN"\GOTO 100
170 J1=Y1-Y+1
```

```
180 IF J1<1 THEN 210\IF J1>4 THEN 210
190 PRINT A$;Y1;"SALES";S(I1,J1)
200 GOTO 220
210 PRINT "NO RECORDS FOR";Y1
220 PRINT "RERUN - TYPE Y FOR YES, N FOR NO";
230 INPUT A$
240 IF A$="Y" THEN 100
250 DATA 1978
260 DATA "JANUARY",59,62,74,82,"FEBRUARY",51,53,56,60
270 DATA "MARCH",43,46,50,68,"APRIL",41,44,49,66
280 DATA "MAY",39,42,47,63,"JUNE",38,40,45,50
290 DATA "JULY",34,39,45,51,"AUGUST",36,41,49,55
300 DATA "SEPTEMBER",41,50,58,61,"OCTOBER",52,69,77,81
310 DATA "NOVEMBER",60,70,79,92,"DECEMBER",63,76,83,94
320 END
```

Sample run

```
SALES INFORMATION
MONTH ? FEB YEAR ? 1979
MONTH NOT KNOWN
MONTH ? FEBRUARY YEAR ? 1979
FEBRUARY 1979 SALES 53
RERUN - TYPE Y FOR YES, N FOR NO ? Y
MONTH ? JUNE YEAR ? 1927
NO RECORDS FOR 1927
RERUN - TYPE Y FOR YES, N FOR NO ? N
```

Y is the year in which the sales record starts (1978). The double loop from lines 40 to 90 reads the data so that M$(I) is the name of the Ith month and S(I,J) represents the sales in the Ith month of year J where 1978 is counted as year one. The user inputs the month A$ and the year Y1 for which the figures are required. The next loop from lines 120 to 150 checks the month input A$ against the names of the month. When A$ is different from all of these (line 130) the program eventually reaches line 160 to print the fact that the name of the month input has not been found – this has happened on the first line of the sample run. Otherwise a match is found and I1 (line 140) saves the number of the month and then calculates the number of the year J1. Line 180 ensures that J1 lies between one and four inclusive; if not, no record exists for the year input (line 210). When a rerun is not wanted, execution will move to line 250, a DATA statement, but the computer will ignore this and the following lines because these statements do not perform any function by themselves.

COMPARING STRINGS

The relation (or order) operators may be used to compare strings according to a numerical code which assigns a number to each alphanumeric character. The code is called the American Standards Code for Information Interchange or **ASCII** for short. It is mainly used to store characters in the computer and send signals to teletypes, line printers and other input and output devices. The coding and decoding is performed automatically by these devices without any specific instructions by the programmer.

Character	Code	Character	Code	Character	Code
space	32	5	53	J	74
!	33	6	54	K	75
"	34	7	55	L	76
#	35	8	56	M	77
$	36	9	57	N	78
%	37	:	58	O	79
&	38	;	59	P	80
'	39	<	60	Q	81
(	40	=	61	R	82
)	41	>	62	S	83
*	42	?	63	T	84
+	43	@	64	U	85
,	44	A	65	V	86
–	45	B	66	W	87
.	46	C	67	X	88
/	47	D	68	Y	89
Ø	48	E	69	Z	90
1	49	F	70	[	91
2	50	G	71	\	92
3	51	H	72	]	93
4	52	I	73	↑	94
				–	95

Some computers have more characters than the ones mentioned above so that the list is incomplete although adequate for most purposes. Note also that many BASIC systems may use a variation of the above code but, in such cases, the functions based on the code operate in exactly the same manner.

The **relation operators** then have the following meaning:

<	earlier in ASCII order than
>	later in ASCII order than
<=	earlier or the same as
>=	later or the same as
=	same ASCII order as (or simply the strings are the same)
<>	different ASCII order than (or simply the strings are different)

For example:

"AND"<"BRAIN" because the first character in AND has a lower ASCII value than the first in BRAIN.
"BLADE">"BLACK" because the first three characters in each string are the same but the fourth in BLADE has a higher ASCII value than the fourth in BLACK.
"SENDING">"SEND" with this code because, for this comparison, the word SEND is treated as having three blank spaces at the end to ensure that the strings are of equal

length. Then the ASCII code for I is greater than the blank space code.

To sort words into alphabetic order, the program on page 44 could be modified by simply replacing every occurrence of A() by A$(), changing D to D$ and rewriting the dimension statement and a few print statements.

STRING CONCATENATION

Strings may be joined together in a single string by means of the **ampersand sign**.

```
10 DIM X$(11)
20 A$="AMPER"
30 B$="SAND"
40 X$=A$&B$
50 PRINT X$
60 END
```

Line 20 assigns the string AMPER to A$ and line 30 assigns SAND to B$. On line 40, X$ becomes AMPERSAND. In many systems a plus (+) sign is used in place of the AMPERSAND.

STRING FUNCTIONS

String handling functions are available to allow the processing of alphanumeric data. The names are not standardized but functions with the following capabilities are generally avoidable.

LEN function This gives the number of characters in a string.
A$="ELEPHANT", LEN(A$)=8
B$="COMPUTER PROGRAM", LEN(B$)=16 (the space is included)
C$="A+B ", LEN (C$=5 (spaces before and after)

ASC function This converts a one character string to its ASCII value.
ASC("A")=65, ASC("3")=51, ASC(" ")=32 (a blank space)

CHR$ function This converts a number to its equivalent ASCII character (the opposite of the ASC function).
CHR$(69)="E", CHR$(36)="$", CHR$(56)="8"
The CHR$ function gives a string – hence the dollar sign.

VAL function This converts a string containing a number into the actual number.
VAL("11")=11, VAL("345.2")=345.2
VAL is a numerical function with a string as its **argument**, i.e. it acts on a string to produce a number.

STR$ function The opposite of the VAL function. It converts a number to its string representation.
STR$(12)="12", STR$(9.62)="9.62"

POS function This has the general form POS(A$,B$,C) and returns the location in the string A$ of the first occurrence of the string B$ starting at the Cth character in A$, e.g.
If A$="JANUARY", B$="A" then
POS(A$,B$,1)=2, i.e. the letter A occurs first in the word JANUARY in position 2.
POS(A$,B$,3)=5, i.e. the next letter A starting from the third character is in position 5 in the word JANUARY.
POS(A$,B$,6)=0 because the letter A does not occur in A$ from the sixth character onward.
If A$="ARITHMETIC", B$="THM"
POS(A$,B$,1)=4, i.e. the string B$ starts in A$ in position 4.
Similarly POS(A$,B$,4)=4 and POS(A$,B$,5)=0.

SEG$ function This has the form SEG$(A$,P,Q) and returns that part of the string A$ between positions P and Q inclusive, e.g.
A$="NEANDERTAL"
SEG$(A$,3,3)="A" and SEG$(A$,3,5)="AND".

In many systems LEFT$, RIGHT$, and MID$ are provided rather than SEG$.

STRING PROCESSING

These functions can be used to perform various typesetting and editing services in data processing and information retrieval systems. As a simple illustration of some of these functions the next program will process a string such as SMITH JOHN L and produce the name as J L SMITH ESQ as it would appear in an address. The blank spaces in the first string are necessary to indicate the start and end of a name or initial, though the program will also take account of extra blanks inserted in the string by mistake.

```
10  DIM N$(50),S$(25),A$(15)
20  PRINT "ADDRESS NAMES"
30  B$=CHR$(32)\A$=B$
40  READ N$\L=LEN(N$)
50  K=0\P=0
60  IF P>=L THEN 150
70  Q=POS(N$,B$,P+1)
80  IF Q<>0 THEN 90\Q=L+1
90  IF Q=P+1 THEN 140
100 K=K+1\IF K=1 THEN 130
110 I$=SEG$(N$,P+1,P+1)&B$
120 A$=A$&I$\GOTO 140
130 S$=SEG$(N$,P+1,Q-1)&B$
140 P=Q\GOTO 60
150 N$=A$&S$&"ESQ"
160 PRINT N$
170 DATA "SMITH JOHN L"
180 END
```

Line 30 defines B$ as the character with ASCII code 32, i.e. a blank space. Alternatively B$ could be replaced by " " whenever it occurs later in the program. A$ is to be the set of initials and initially contains the **null** or **empty string**, i.e. a blank space.

Line 40 The string N$ is read and its length L evaluated.
Line 50 K is used solely to determine the surname on line 100. P represents the position of the first space and takes the value zero initially.
Line 60 A check is made to see if the string has been fully analyzed.
Line 70 Q is the position of the next space after position P.
Line 80 When Q is zero there are no more spaces and Q is set to L+1 to represent an artificial space at the end of N$.
Line 90 When Q=P+1 the spaces are consecutive and are ignored.
Line 100 K is incremented by 1.
Line 110 I$ represents the initial in position P+1 with a blank after it.
Line 120 I$ is placed at the end of the set of initials A$.
Line 130 K must be 1 so the whole surname is stored in S$ with a blank after it.
Line 140 P is reset to the value of Q and the process repeated.
Line 150 N$ becomes the name for the address.

Dry run

Suppose N$ has the form

Position	1	2	3	4	5	6	7	8	9	10	11	12	13	14	15	16
Character			S	M	I	T	H		J	O	H	N			L	

where positions 1, 2, 8, 13, 14 and 16 represent the blank spaces.
Then L = 16, K=0, A$="".

P	**Q**	**Q=P+1**	**K**	
0	1	Yes	–	
1	2	Yes	–	
2	8	No	1	S$=SEG$(N$,3,7)&B$="SMITH "
8	13	No	2	I$=SEG$(N$,9,9)&B$="J " A$=A$&I$=" J "
13	14	Yes	–	
14	16	No	3	I$=SEG$(N$,15,15)&B$="L " A$=A$&I$=" J L "

16 Analysis complete because P>=L.
Then N$ becomes " J L SMITH ESQ"
If necessary the space at the beginning of N$ could be removed by the statement
155 N$=SEG$(N$,2,LEN(N$)).

Chapter 10

Avoiding Errors

A program rarely works as intended on its first run on the computer. Either an **error code** is generated indicating that the program does not conform to the BASIC language or the output bears little relation to the correct answer. The types of mistakes may be classified into two distinct groups: **syntax errors** which are errors in language, and **semantic errors** which are errors in meaning or logic. The former will always be detected by the computer but the latter must be found by the programmer or user.

Before a computer can run a program in BASIC or any other high level language such as Fortran, Cobol or Pascal, it has to be translated into a machine language, i.e. a series of instructions written in binary code. The program which performs the translation is called a **compiler** (or an **interpreter**, which functions in a slightly different way) and is an extremely complex program stored in the computer. When the BASIC program is used, the compiler automatically performs this translation and then runs the program. Syntax errors will be detected during the compilation (translation) or the running of the program. Typical errors encountered during compilation are incorrect typing (FOT instead of FOR), a missing END statement, missing equal sign or bracket, a FOR loop without a corresponding NEXT statement, an undefined function, program too large for memory or a GOTO statement to a line number which does not appear in the program. The computer will print a code for these errors, e.g. ME means missing END statement, and these may be looked up in a table of **compile time diagnostics** which gives the meaning of each error code. Errors detected during the run are also listed in a table of **run time diagnostics**. Typical errors here are attempting to read beyond a data list, not dimensioning a variable or not allocating enough memory for such a variable, incorrect data input such as trying to store a word in a numeric variable, attempting to find the logarithm of a negative number or attempting to divide by zero. Both types of syntax error are usually very minor and can easily be corrected by retyping the necessary lines of the program.

Semantic errors will not be detected by the compiler as they do not contravene the laws of the language. They can be relatively minor – typing a plus sign instead of a minus sign, forgetting to put a comma or semicolon in a PRINT statement, transferring to the wrong line number or omitting a statement completely. Serious faults occur when the logic or design of the program is unreliable. Semantic errors can be extremely difficult to detect and the time taken in correcting a program is often greater than the time needed to write it.

Methods of keeping errors to a minimum will now be considered.

PROGRAM DESIGN

The task of constructing a computer-based solution to a problem can be broadly classified under the following headings:
Defining the problem
Constructing a solution
Flow charting
Coding
Constructing test data
Debugging
Documenting

Defining the problem

The programmer must develop a clear definition of the problem and decide if a computer solution is in fact appropriate. Sometimes the use of mental arithmetic or an electronic calculator will be most appropriate as in many of the programs of Chapter 2. The definition of the problem should include the scope and generality to be achieved and the type of input and output required.

Constructing a solution

Before considering the programming techniques that might be needed, the programmer has to determine the type of solution to be applied to all or part of the problem. Many statistical methods such as regression analysis have already been developed as a set of formulae and decision processes. An algorithm may exist to solve the problem, e.g. the Simplex Method or one of its variations in linear programming problems. Otherwise the programmer may have to resort to a **heuristic** approach, i.e. a set of principles which often leads to the correct solution or a solution close to the optimum but is not completely foolproof. When no known algorithm or heuristic exists or their implementation would exceed computer memory or take a prodigious length of time to run, the problem may be best solved by **Monte Carlo methods** (any technique using random numbers as in queuing simulations), numerical methods (techniques leading to an approximate solution such as **iterative processes**) or by a trial and error method as the last resort.

Flow charting

The structure or layout of the program can now be decided upon. In the case of a program to determine the median of a set of numbers, the structure could be exhibited as follows:

1. Input the numbers and frequencies.
2. Find the cumulative frequencies.
3. Sort the numbers into order, swapping the cumulative frequencies if necessary.
4. Choose the middle number taking the cumulative frequencies into account.
5. Print the result.

The program would perform each individual task in the order shown above. When greater detail is needed for all or part of the program, a flow chart can be constructed to show clearly the logic behind each process. For example a flow chart to perform the addition or deletion of a name using the list A$(1), . . . , A$(N). Here additions are made to the end of the list and deletions are made from the front – an example of a **queue** or **push up list**. The first name on the list is the first one out (**FIFO** – first in, first out). The list contains N names originally and the variable P is given the value N+1. P is called a **pointer** and stores the location in which the next addition is to be placed. It is assumed that the maximum permissible queue length is M.

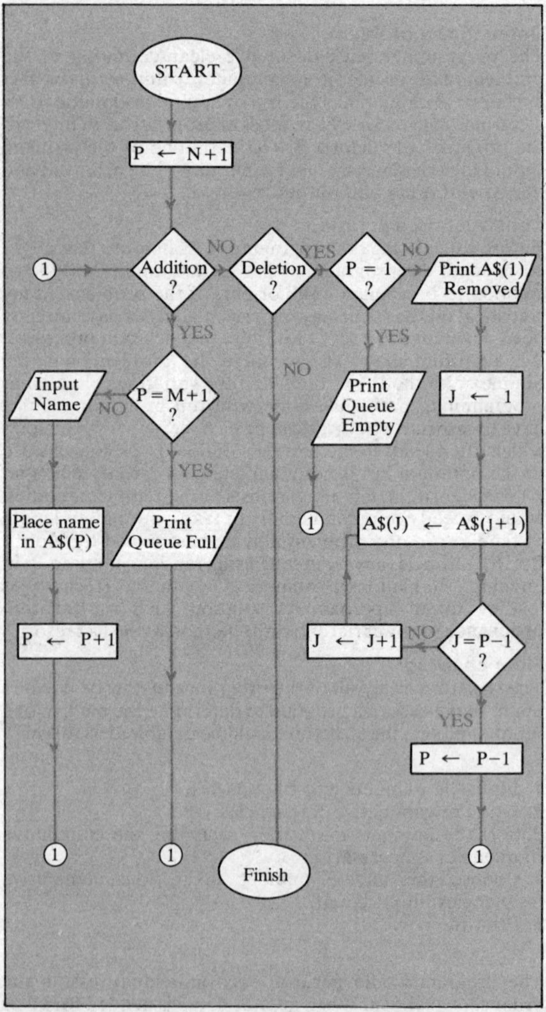

The following conventions have been used:

Ovals	the start or finish of the flow chart
Rectangles	assignment statements
Diamonds	**decision boxes** containing questions
Parallelograms	input and output statements
Circles	**connector symbols**
J←1	place the number one in store J (J=1)
Arrows	the direction of flow of the lines

To remove the first name on the list, the names are each shifted to the next location down (in a numerical sense) by the J loop on the right of the flow chart. In this way the name to be deleted is lost and the name in location P−1 becomes the null string. The pointer is then reduced by one before returning, via the connector symbol, to the next addition, deletion or stop sequence. The flow chart is much simpler in the case of a **stack** (or **push down list**) when additions and deletions are both made at the same end. Here the last name in is the first one out (**LIFO** for short). A flow chart is much easier to follow than a program and makes the actual programming itself quite a simple translation process. A complex problem requiring a long program (or a set of programs) will be broken down into many flow charts with connector symbols joining each one.

Coding

This process is the writing of the program in an appropriate computer language. If flow charts have already been developed in considerable detail, the coding will largely be a mechanical task although certain pointers towards producing a good programming style are worth mentioning. In order to ensure that the program is easy to read for any later corrections, the programmer can insert comments to indicate the function of a particular part of the program. This is achieved by the **REM statement**, e.g.

```
50 REM THIS PART CALCULATES THE CORRELATION
```

This line is ignored by the compiler and does not perform any actual programming function. It is advisable to avoid long complicated formulae by breaking them down into several statements. One instruction per line will make the program easier to follow and correct, though this practice has not always been adhered to in this book in order to save space. The allocation of loop variables as I, J, K, and summation variables as R, S, T and the use of C to represent the cost, where possible and appropriate, will make it easier to recognize the purpose of a variable.

Incorrect input of the data can easily cause a program to produce odd results. When an input is supposed to be a whole number in the range 1 to 8 say, a check can be made within the program to only accept the correct type of input:

```
50  PRINT "NO. YEARS–AN INTEGER FROM 1 TO 8 INCLUSIVE"
60  INPUT N
70  IF N<1 THEN 50\IF N>8 THEN 50
80  IF N<>INT(N) THEN 50
```

Here line 50 prints out exactly the kind of input expected and the following lines ensure a return for an acceptable input if necessary, Alternatively a PRINT statement could be used to inform the user that the input is unacceptable. When a long list is to be input it is easy to type in an incorrect number and then have to rerun the program and start all over again. The difficulty may be avoided as follows:

```
40  N=0
50  PRINT "INPUT THE SALES-TYPE -1 TO FINISH"
60  PRINT "PRESS THE RETURN KEY AFTER EACH INPUT"
70  N=N+1
80  PRINT "SALES FOR MONTH NUMBER";N;
90  INPUT S(N)
100 PRINT "IS THIS CORRECT-TYPE Y FOR YES, N FOR NO";
110 INPUT A$
120 IS A$<>"Y" THEN 80
130 IF S(N)<>-1 THEN 70
140 N=N-1
```

When a quantity is input incorrectly the user has the chance to change that input by typing the letter N as answer to the question on line 100 – in this case the variable N, counting the number of sales input so far, is not incremented. Line 140 reduces N by one after the list has been completed because the last input was the rogue value. When the program contains too much repetition of code, the use of subroutines (see next chapter) or the redesigning of parts of the program should be considered. The number of lines required can be reduced by rewriting sections of the program in a neater fashion, e.g.

```
50 IF A>12 THEN 70
60 GOTO 80
70 A=0
80 . . . . . . .
```

can be replaced by

```
50 IF A<=12 THEN 70
60 A=0
70 . . . . . . .
```

Both sets of code set A to zero if it is larger than 12 and leave it as it is otherwise, but the second uses fewer transfer statements.

Exit from a loop can cause some problems when the value of the loop variable on exit is needed, e.g.

```
110 FOR I=1 TO 5
120 IF A(I)=3 THEN 160
130 NEXT I
140 PRINT "VALUE NOT FOUND"
150 GOTO 170
160 PRINT "VALUE IN PLACE";I
170 . . . . . . . . .
```

```
110 FOR I=1 TO 5
120 IF A(I)<>3 THEN 140
130 I1=I\GOTO 170
140 NEXT I
150 PRINT "VALUE NOT FOUND"
160 GOTO 180
170 PRINT "VALUE IN PLACE";I1
```

Both of these subprograms search for the value three among the variables A(1), . . . , A(5). The first leaves the loop without storing the value of the variable I before exit. The second stores I in I1 before exit and is to be preferred despite it using two statements more than the first. Finally, consider the queuing simulation on page 57 where the two variables Z1 and Z2 are given the values 5 and 4 respectively on line 40. These values are used later in the program but only mentioned implicitly as Z1 and Z2. They are called **parameters** and make the task of altering their values considerably easier. If a program contains any numbers as in FOR I=1 TO 5, it is sometimes worthwhile to change the 5 (for example) into a parameter A by adding the line A=5 near the beginning of the program and changing the FOR statement to FOR I=1 to A.

Constructing test data

The testing of a program is possibly the most important aspect of the whole design process. Any possible errors might be encountered during the flow charting or coding stages and the use of dry runs at any stage can be most helpful in keeping semantic errors to a minimum. Unfortunately many programs are so complex that systematic checking by hand is a long and tedious business and errors are almost certain to be present after coding. To test the program it must be run on the computer with various possible inputs for which the answer is preferably already known or, with a very long program, each section of it may be run and tested separately. The test data must include any special cases and be sufficient in scope to check invalid inputs and all parts of the program.

Debugging

This is the process of correcting errors within the program. There is no guaranteed method of locating faults but certain checks and aids will help in the search. Syntax errors usually present no problem because of the generated error codes although some run time errors can be difficult to detect. To isolate a fault, additional PRINT and STOP statements can be temporarily added to the program and a check made of the value of the variables after each stage. It might be helpful to perform a dry run on all or part of the program using some of the test data and the results compared with the computer printout. Tracing the path through a program may be achieved using the **TRACE function** as follows:

```
10 X=3\Y=2\Z=X+Y\A(0)=-1
20 FOR I=1 TO X\READ A(I)\NEXT I
30 FOR I=1 TO Y\READ B(I)\NEXT I
35 TRACE ON
40 IF A(X)>B(Y) THEN 60
50 A(Z)=B(Y)\Y=Y-1\GOTO 70
60 A(Z)=A(X)\X=X-1
70 Z=Z-1\IF Y>0 THEN 40
75 TRACE OFF
80 DATA 2,4,5,1,3
90 . . . . . . . .
```

Computer Output

```
% 40 %      % 40 %
% 60 %      % 60 %
% 70 %      % 70 %
% 40 %      % 40 %
% 60 %      % 50 %
% 70 %      % 70 %
% 40 %      % 75 %
% 50 %
% 70 %
```

This program will **merge** the ordered lists A(I) and B(I), containing positive numbers, into a single list in such a way that the final answer is stored in A(I) (it is more efficient to store the final list in the longer of the two initial lists). X, Y and Z are pointers to indicate the position of the numbers in A(I), B(I) and the final list respectively. The two extra lines, 35 and 75, are added to the program to trace the order of line execution between these two line numbers. TRACE ON switches the function on and TRACE OFF switches it off. Each line number executed when the trace is on has been printed in between percentage signs to distinguish it from other output which might be printed by the main program. The line numbers are in fact printed in a single column by the computer.

Faults sometimes occur due to the small inaccuracies of the computer. A calculation which should give the answer X=0 may in fact give X=0.0000009 so that a conditional statement such as IF X=Ø THEN 9Ø will fail to function as desired. The problem can be solved by replacing the above by IF ABS(X)<Ø.ØØØØØ1 THEN 9Ø. Also a program which works perfectly but takes too long to run may have to be rewritten in order to save time on the computer. Sometimes the program will never finish because of an **infinite loop**, e.g.

```
3Ø  N=N+1
4Ø  GOTO 3Ø
```

These two lines will be executed indefinitely because there is no exit from the loop. This is an extremely simple case – an infinite loop can be very difficult to detect when it involves the whole program. After the program changes, the test data must be thoroughly tested again until the output seems to be satisfactory. Even then one can never be absolutely certain that the program will work perfectly in the future.

Documenting

It is also desirable to keep a record of the program for anyone who might wish to use or adapt the program at a future date. This is particularly important where a program library will be set up. The record should include the definition of the problem, the solution and its scope, relevant flow charts, a definition of the variables used, a computer printout of the program, the test data and any relevant facts which a user should be aware of before running the program.

Chapter 11

Subroutines

Within certain programs it is possible that a calculation has to be performed at many stages during execution. For example it may be necessary to check all inputs to ensure that they are all positive whole numbers less than a given number. It would be bad programming to write the same sort of check after every input. A neater method is to use **subroutines** as follows:

```
10  INPUT A\N=A
20  M=12\GOSUB 500
30  IF F=1 THEN 10
40  INPUT B\N=B
50  M=20\GOSUB 500
60  IF F=1 THEN 40
70  FOR I=1 TO 5
80  INPUT C(I)\N=C(I)
90  M=50\GOSUB 500
100 IF F=1 THEN 80
110 NEXT I
```

Lines 120 to 480 contain the main part of the program

```
490 STOP
500 F=0
510 IF N<>INT(N) THEN 550
520 IF N<0 THEN 550
530 IF N>M THEN 550
540 GOTO 580
550 F=1
560 PRINT "INPUT WHOLE NUMBER"
570 PRINT "BETWEEN 0 AND";M;"INCLUSIVE"
580 RETURN
590 END
```

A is input on line 10, its value stored in N and M and the maximum permissible value of A set to 12. The instruction **GOSUB 500** means transfer to the subroutine between lines 500 and 580 inclusive. The subroutine sets the flag F to one when N is not a whole number, is negative or when it is larger than M and then prints the fact that the input is inadmissible. Otherwise the flag remains as zero (line 500). When the **RETURN statement** on line 580 is executed the program will automatically transfer to the line after the original GOSUB command, i.e. line 30 in this case. Here the flag is checked to see whether another input is required and if so the whole process will be repeated. The next input on line 40 is treated similarly (B must be less than or equal to 20) but the RETURN statement in the subroutine returns control to line 60 because this is the line after the last GOSUB call. The subroutine can also be used inside a loop – there is no problem here of jumping back into the loop. Line 490 at the end of the main part of the program ensures that the subroutine will not be executed without a GOSUB call. The variables N and M are necessary solely for the subroutine which has to check different inputs. Note that a combination of GOTO statements cannot replace the subroutine because of the unique ability of a subroutine to return control to varying lines of the main program. A program may contain a number of subroutines and it is common practice to place them at the end of the program before any DATA statements. Exit from a subroutine must always be achieved by the RETURN statement. A subroutine may call another subroutine – a process known as **nested subroutines**. In this case the line numbers to be returned to are placed in a stack by the compiler, i.e. return is always made first to the line after the last GOSUB call. Some programmers use subroutines in the following manner:

```
10  Declaration of variables       100–190  Input subroutine
20  REM INPUT THE DATA             200–390  Calculation subroutine
30  GOSUB 100                      400–490  Print subroutine
40  REM MAIN PROGRAM               500–690  Other subroutines
50  GOSUB 200                               called by above
60  REM PRINT THE RESULTS          700–790  Data statements
70  GOSUB 400                      800      END
80  STOP
```

This is an example of a **top down design** and makes the program easier to check although many subroutines are only executed once and are not absolutely necessary. Subroutines are very useful when results need to be printed in a certain form, e.g. all answers to two decimal places.

SOLVING EQUATIONS

A project has the following cash flows:

	Year 0	Year 1	Year 2	Year 3	Year 4	Year 5
$	−5000	1600	2000	−3500	2500	6000

The internal rate of return (IRR) is the discount rate for which the net present value (NPV) is zero. It is impossible to find the IRR by an algebraic method although it can be estimated to a high degree of accuracy by a standard technique employing a binary search. The first task is to find two discount rates R1 and R2 whose NPVs are negative and positive respectively. The IRR must be between these two values and the midpoint R of this range is considered next. If R has a negative NPV the IRR is in the range R to R2, if it has a positive value, the IRR is in the range R1 to R. The interval is halved at each stage and the method continues until this interval is extremely small giving a very good approximation to the IRR. A subroutine is used to calculate the NPV from the discounted cash flows.

```
10 PRINT "INTERNAL RATE OF RETURN"
20 DIM C(25),A$(15)
30 PRINT "INPUT THE CASH FLOWS"
40 PRINT "TYPE E TO FINISH"
```

```
50  Y=0\F=0
60  PRINT "YEAR";Y;\INPUT A$
70  IF A$="E" THEN 100
80  C(Y)=VAL(A$)
90  Y=Y+1\GOTO 60
```

The cash flow is input as a string and the string changed to a number using the VAL function when necessary.

```
100  PRINT "SPECIFY THE RANGE OF DISCOUNT RATES"
110  PRINT "LOWER LIMIT";\INPUT A
120  PRINT "UPPER LIMIT";\INPUT B
```

The user is allowed to specify the range of discount rates to be considered in case negative rates are irrelevant.

```
130  REM FIND STARTING VALUES
140  R=A\GOSUB 430
150  IF S=0 THEN 360
160  R1=R\S1=S
170  R=R+1
180  IF R>B THEN 400
190  GOSUB 430
200  IF S=0 THEN 360
210  R2=R\S2=S
220  IF S1<>S2 THEN 240
230  R1=R2\GOTO 170
```

The subroutine called takes a discount rate R and returns S, the sign of the NPV. Line 140 finds S for the rate A and, if this is zero, the IRR has been found immediately. Otherwise the rate is saved in R1 and the sign of the NPV saved in S1. R is increased by one and when it is larger than B the program ends. A similar process follows for this new rate but the answers are stored in R2, S2. When the condition on line 220 is satisfied, two rates with NPVs of differing signs have been found and the binary search can proceed. In the case of this condition failing (line 230), R1 is reset to R2 (the latest rate under consideration), S1 need not be altered because it must have the same value as S2 and transfer is made to line 170 to consider the next rate. In this way the binary search always starts with two rates differing by one. This method is not the quickest (a random search can be used to find rates with differing signs) but will allow the program to find multiple internal rates of return in unconventional cash flow projects. The addition of one to find the next rate on line 170 is somewhat arbitrary and means that the program will not detect IRRs within 1% of each other. There is no fool-proof method of solving equations unless every possible rate is checked separately.

```
240  REM BINARY SEARCH
250  IF S1=-1 THEN 270
260  D=R1\R1=R2\R2=D
270  REM R1 IS NOW RATE WITH NEGATIVE NPV
280  R=(R1+R2)/2
290  GOSUB 430
```

```
300 IF S=0 THEN 360
310 IF S=1 THEN 330
320 R1=R\GOTO 340
330 R2=R
340 IF ABS(R1-R2)<=0.000005 THEN 360
350 GOTO 280
```

R1 and R2 are interchanged if necessary and the midpoint R of the range evaluated. When S is zero the IRR has been found, when S=1 its NPV is positive and R2 is replaced by R and when S=−1 its NPV is negative and R1 is replaced by R. When R1 and R2 are very close together the answer is printed immediately (line 340).

```
360 REM PRINT THE ANSWERS
370 F=1
380 PRINT "IRR IS";R
390 A=INT(R+1)\GOTO 140
400 IF F=1 THEN 420
410 PRINT "NO IRR FOUND"
420 STOP
```

F is a flag variable – when it is one an answer has been found. Line 390 will carry on the search for any more answers to the problem. The next line is reached via line 180.

```
430 REM SUBROUTINE
440 S=0
450 D=1+R/100
460 FOR I=0 TO Y-1
470 S=S+C(I)/D↑I
480 NEXT I
490 S=SGN(S)
500 RETURN
510 END
```

D is the discount rate and Y−1 is the final year for which a cash flow has been input (see lines 60, 70).

Sample run for example shown

```
INTERNAL RATE OF RETURN
INPUT THE CASH FLOWS
TYPE E TO FINISH
YEAR 0 ? -5000
YEAR 1 ?  1600
YEAR 2 ?  2000
YEAR 3 ? -3500
YEAR 4 ?  2500
YEAR 5 ?  6000
YEAR 6 ?  E
SPECIFY THE RANGE OF DISCOUNT RATES
LOWER LIMIT ? 0
UPPER LIMIT ? 100
IRR IS 15.0936
```

Chapter 12

Data files

The input of data to a program can be a very long and time-wasting process, especially when each run requires virtually the same information. The use of DATA statements avoids this problem but reduces the scope of the program. Sometimes the data or results of one program are needed in another program and transferring the data by hand is not always practical. A solution to these problems is to store information in the computer or one of its peripherals using **a data file**. A file is a list of data usually stored on magnetic tapes, floppy discs, etc. The only type of file considered is a **serial file**, i.e. a list of data whose contents can only be accessed by reading from the beginning of the file. In a **random access file**, for example, the reading of data can begin in any position. The operations and mnemonics used in file processing are far from being standardized in BASIC but the general ideas behind file handling can be understood by making certain assumptions. A **numeric file** contains numerical data, an **alphanumeric file** contains alphanumeric data and a file may not contain both types of information. An **input file** is used to read data from, an **output file** is used to write data into and a file may not be employed as an input file and an output file at the same time. Sometimes it is necessary to indicate the device on which the file is stored, the type of data stored in it, and its status (input or output) but these are very minor details and are omitted in this chapter.

The set of file instructions is as follows:

OPEN #1, "TOWNS"	A file must be described and opened before use. This statement opens a file called TOWNS which will be referred to as file number 1 in the program.
OPEN #2, "TOWNS"	This opens the same file TOWNS but gives it the number 2 instead.

The number of the file is always placed after the # (sharp) symbol and the name of the file enclosed in quotation marks after the colon. It will be assumed that the name may have a maximum of six alphanumeric characters.

INPUT#1,A	Read the next number in file number 1 and place it in store A.
INPUT#2,A$(I)	Read the next word in file number 2 and place it in the subscripted string variable A$(I).
PRINT#1,A	Write (i.e. place) the contents of variable A into file number 1.
PRINT#3,S(7)	Write the contents of the subscripted variable S(7) into file number 3.
RESTORE#1	Reset the pointer to the beginning of file number 1, i.e. reading of

	data starts from the first number or word in the file – compare the RESTORE statement as used with DATA statements.
SCRATCH#1	Delete all the data in file number 1.
IF END#1 THEN 70	If the end of the file has been reached, transfer to line 70.
CLOSE#1	Close the file, i.e. file instructions cannot be performed on the file unless it is re-opened.

All files must be closed after use, otherwise the contents might be lost or deleted. One or more files may be open at the same time but they must have a different number in order to avoid ambiguity in any of the above file operations. Of course once a file has been closed, its number may then be used to open another file. Sometimes there is a restriction on the number of files open at any time, for example, the maximum might be four files numbered 1, 2, 3, 4. A file may also be opened using the statement OPEN 1,A$. Here the name of the file is assumed to be already stored in A$ and the number in the variable I. The operations of reading, creating, copying and updating files will now be illustrated.

CREATING A FILE

To set up a numeric file to be used and updated by other programs.

```
10  PRINT "FILE NAME"          The name of the file is input
20  INPUT A$                   at the keyboard, e.g. DATA1.
30  OPEN#1,A$                  The file is described and opened.
40  PRINT "INPUT THE DATA"
50  INPUT B                    The next number is input at the
60  IF B= -1 THEN 90           keyboard. -1 is the rogue value.
70  PRINT#1,B                  The number is written into the file.
80  GOTO 50                    Return for next input.
90  CLOSE#1                    Close the file.
100 END
```

If the numbers input are 5, 2, 3, 7, −1, the file called DATA1 will contain 5, 2, 3, 7 in that order. An extra line, 35 SCRATCH#1, can be added in case the file already exists and its contents are to be deleted first. An example of the creation of a file within a larger program will be illustrated on page 100.

COPYING A FILE

Many computer departments keep a second or third copy of a file as a protection against accidental deletion. The next example makes a copy of an existing file called STAFF.

```
10  PRINT "FILE NAME"
20  INPUT A$                   File name is input, e.g. STAFF.
30  B$=A$+"C"                  The copy is to be called STAFFC.
```

40 OPEN#,A$	STAFF is opened as an input file.
50 OPEN#2,B$	STAFFC is opened as an output file.
60 IF END#1 THEN 100	IF the end of STAFF has been reached.
70 INPUT#1,C$	goto 100. Read the first name in STAFF.
80 PRINT#2,C$	The name is written into STAFFC.
90 GOTO 60	Return to end of file test.
100 CLOSE#1\CLOSE#2	Close both files.
110 END	

The end of file test is very important because attempts to read beyond the end of a file can result in the loss of the file. In some versions of BASIC, this test must be placed after the INPUT# or PRINT# statement and indicates that the last attempt to read or write was unsuccessful. The reading and writing of data for files is analogous to inputting and printing data at the keyboard. Usually a special statement, LPRINT, is used to send output to a printer. The syntax is the same as the PRINT command. The syntax for file input and output varies widely between different computers. The reader should consult his BASIC manual for the exact specification of these statements.

UPDATING A FILE

The next example reads an array from a file and allows the user to update it. The results of a questionnaire developed by a Market Research Department are already stored in the file as 4, 3, 2, 1, 3, 3, 1, 2, 3, 1, 2, 2, 3, 1. This information is to be placed in an array as follows:

Choice number

		1	2	3
	1	2	1	3
Person	2	3	1	2
	3	3	1	2
	4	2	3	1

Four people have been asked to rank three products in order of preference, e.g. person two has placed product three first in his choice list.

The first number in the file is not part of the array. It is used as a **key** to indicate the number of people whose results have been stored. The second number gives the number of products.

```
10 DIM P(100,5),C(100,5),A(5,6)
20 OPEN 1,"MRQ1"
30 INPUT #1,MN\IF M=O THEN 70
40 FOR I=1 TO M\FOR J=1 TO N
50 INPUT #1,P(I,J)
60 NEXT J\NEXT I
```

The number of people M and the number of products N are read on line 30. An end of file test is not needed here because the number of further entries in the file is now determined by the two key values. The array is then read P(I,J) storing the product placed in position J by person I.

```
70 PRINT "ADD MORE RESULTS"
80 PRINT "ANOTHER PERSON";\INPUT A$
90 IF SEG$(A$,1,1)="N" THEN 170
100 PRINT "GIVE PRODUCT NUMBERS"
110 M=M+1
120 FOR J=1 TO N
130 PRINT "CHOICE";J;"PRODUCT";
140 INPUT P(M,J)
150 NEXT J
160 GOTO 80
```

Line 90 only checks the first letter of the input to see if the word input is YES or NO (Y or N will also be treated in the same way). The new results are now stored in the array with the old results but the file contents have not been altered. The analysis of the array is now performed.

```
170 FOR P=1 TO N\FOR J=1 TO N+1
180 A(P,J)=0
190 NEXT J\NEXT P
200 FOR I=1 TO M\FOR J=1 TO N
210 P=P(I,J)\C(I,P)=J
220 A(P,J)=A(P,J)+1\A(P,N+1)=A(P,N+1)+J
230 NEXT J\NEXT I
```

I is the person, J the choice number and P the product. The array C(I,P) gives the choice number of product P for each person I and presents the results in a different form. A(P,J) counts the number of times that product P is placed in position J and A(P,N+1) stores the sum of all the choice numbers for each product. Note that C can be set up using the statement C(I,P(I,J))=J. Here the variable P is not needed and could be similarly replaced on line 220. The results are now printed.

```
240 PRINT "CHOICE NUMBERS"
250 FOR I=1 TO 15\PRINT"-";\NEXT I
260 PRINT\PRINT
270 PRINT "PERSON";
280 FOR P=1 TO N\PRINT TAB(10*P);"PRODUCT";P;\NEXT P
290 PRINT
300 FOR I=1 TO M\PRINT TAB(3);I;
310 FOR P=1 TO N\PRINT TAB(10*P+3);C(I,P);\NEXT P
320 PRINT
330 NEXT I
340 PRINT
350 PRINT "SUMMARY BY PRODUCTS"
360 FOR I=1 TO 20\PRINT"-";\NEXT I
370 PRINT\PRINT
380 PRINT "PRODUCT";
390 FOR J=1 TO N\PRINT TAB(10*J);"CHOICE";J;\NEXT J
400 PRINT TAB(10*N+10);"TOTAL"
410 FOR P=1 TO N\PRINT TAB(4);P;
420 FOR J=1 TO N\PRINT TAB(10*J+3);A(P,J);\NEXT J
430 PRINT TAB(10*N+12);A(P,N+1)
440 NEXT P
```

Sample output (only using original table contents)

CHOICE NUMBERS

PERSON	PRODUCT 1	PRODUCT 2	PRODUCT 3
1	2	1	3
2	2	3	1
3	2	3	1
4	3	1	2

SUMMARY BY PRODUCTS

PRODUCT	CHOICE 1	CHOICE 2	CHOICE 3	TOTAL
1	0	3	1	9
2	2	0	2	8
3	2	1	1	7

From the last table, it can be seen that product 3 appears to be preferred because it has the lowest choice total. The last part of the program stores the new results in the file. The file is now to be used as an output file and was initially opened as an input file. One way of avoiding this difficulty is to close the file and re-open it as an output file.

```
450 CLOSE#1
460 FILE#1,"MRQ1"
470 SCRATCH#1
480 PRINT#1,M,N
490 FOR I=1 TO M\FOR J=1 TO N
500 PRINT#1,P(I,J)
510 NEXT J\NEXT I
520 CLOSE#1
530 END
```

The old contents of the file are deleted on line 470 and the whole array P(I,J), including the original data, printed back into it. The key variables have to be placed at the beginning of the file so that the format is not altered in any way for any future additions to the file.

TRANSPORTATION

A company has a certain number of depots located in various parts of the country and has received orders for a particular product from its customers. The company wishes to minimize delivery costs by sending the product from the appropriate depot to each customer. This is a standard transportation problem and can be solved using a step by step algorithm. The program uses three files to store the data required. The first contains the current stocks in each depot, the second the current orders from the customers and the third the transportation costs per unit of the product.

```
10 DIM A$(3,6),I(2),N(2),C(25,25),X(25,25)
20 DIM F(25,2),P(25,2),S(25,2),T(50)
30 PRINT "PRODUCT FILE"\INPUT P$
```

```
40 FILE#1,P$
50 FOR I=1 TO 3\INPUT#1,A$(I)\NEXT I
60 CLOSE#1
70 K=1\GOSUB 1000\IF N(1)=0 THEN 970
80 K=2\GOSUB 1000\IF N(2)=0 THEN 950
```

The file opened on line 40 is called a **directory file** and contains the names of the three files used to store the data. The subroutine called on lines 70 and 80 is given below.

```
1000 M=0\N(K)=0\T(K)=0
1010 FILE#K,A$(K)
1020 IF END#K THEN 1070
1030 M=M+1\INPUT#K,Z
1040 IF Z=0 THEN 1020\N(K)=N(K)+1
1050 S(N(K),K)=Z\F(N(K),K)=M
1060 T(K)=T(K)+Z\GOTO 1020
1070 RETURN
```

When K=1, the file containing the depot stocks is opened as file number 1. M counts the number of depots on file, N(1) the number of depots with a non-zero stock level and T(1) the total stocks. For example, suppose that this file contains the entries 15, 0, 6, 32.

M	Z	N(1)			T(1)
0	–	0			0
1	15	1	S(1,1)=15	F(1,1)=1	15
2	0	–			–
3	6	2	S(2,1)= 6	F(2,1)=3	21
4	32	3	S(3,1)=32	F(3,1)=4	53

The second depot has no stock and is omitted from consideration. S(I,1) now contains the stocks for the three remaining depots and F(I,1) stores the positions in the file of these depots. F could be used as a **link** (pointer) to the names of the depots stored in some other file but this is not included here. Similarly when K=2 the following allocations will be made assuming that the customers' orders file contains 0, 7, 10, 8, 13:
M=5, N(2)=4, S(1,2)=7, S(2,2)=10, S(3,2)=8, S(4,2)=13, F(1,2)=2, F(2,2)=3, F(3,2)=4, F(4,2)=5, T(2)=38.
If N(1) is zero or N(2) is zero the program terminates.

```
950 PRINT "NO ORDERS"
960 CLOSE#2\GOTO 980
970 PRINT "NO STOCKS"
980 CLOSE#1\STOP
```

If N(1)=0, only file number 1 has been opened and needs to be closed, otherwise both must be closed. The transportation costs are now extracted from the third file.

```
 90 FOR I=1 TO 25\FOR J=1 TO 25
100 C(I,J)=0\X(I,J)=0
110 NEXT J\NEXT I
```

```
12Ø  FILE#3,A$(3)\Z=Ø
125  FOR I=1 TO N(1)\FOR J=1 TO N(2)
13Ø  Y=M*(F(I,1)-1)+F(J,2)
14Ø  Z=Z+1\INPUT\3,C
15Ø  IF Z<>Y THEN 14Ø
16Ø  C(I,J)=C
17Ø  NEXT J\NEXT I
18Ø  CLOSE#3
```

This file is given the number 3 because two files are still open with numbers 1 and 2. Suppose that this file contains 3, 4, 2, 1, 2, 5, 6, 3, 1, 2, 2, 3, 5, 4, 3, 1, 5, 5, 2, 1. The first five numbers give the transportation costs from the first depot on file to each of the five customers on file, the next five the costs from the second depot on file and so on. Because some depots and customers are not being used, some costs are not relevant to the problem. A **mapping function** Y is used to determine the position in the file of the required costs.

I	J	F(I,1)	F(J,2)	Y	Z	C	
1	1	1	2	5×(1−1)+2=2	1	3	Z<>Y
					2	4	C(1,1)=4

The first relevant cost is that for depot 1 and customer 2 *in the file* and Y gives the position in the third file of this cost. Data are extracted from the file until the value of the counter Z is equal to Y and then the cost saved. Note that C(1,1) refers to the cost for the first depot and first customer *in the program*. The position of the next relevant cost is then determined and the process repeated. The next step is to ensure that the total stock is exactly equal to the total orders – achieved by adding a dummy depot or dummy customer, when necessary, at zero transportation costs.

```
19Ø  F=SGN(T(2)-T(1))\IF F>=Ø THEN 21Ø
2ØØ  F=2
21Ø  N(F)=N(F)+1\S(N(F),F)=ABS(T(1)-T(2))
```

In the present example T(1)=53, T(2)=38 and the flag F becomes −1 on line 19Ø and 2 on line 2ØØ. N(2) is then incremented to 5 and S(5,2) becomes 15, a dummy customer having been added. Similarly when a dummy depot is needed (T(2)>T(1)) the flag becomes 1 and when T(1)=T(2) it takes the value 0. In the latter case, N(0) and S(1,0) are changed on line 21Ø but, as these stores are not used in the program, it is equivalent to not using dummies. The next step is the initial allocation of deliveries X(I,J) from each depot I to each customer J in the program.

```
22Ø  I=1\J=1\C=Ø
23Ø  IF S(I,1)<S(J,2) THEN 26Ø\IF S(I,1)>S(J,2) THEN 28Ø
24Ø  X(I,J)=S(I,1)\X(I,J+1)=Ø.ØØ1\GOSUB 11ØØ
25Ø  I=I+1\J=J+1\GOTO 3ØØ
26Ø  X(I,J)=S(I,1)\S(J,2)=S(J,2)-X(I,J)\GOSUB 11ØØ
27Ø  I=I+1\GOTO 3ØØ
28Ø  X(I,J)=S(J,2)\S(I,1)=S(I,1)-X(I,J)\GOSUB 11ØØ
29Ø  J=J+1
3ØØ  IF I>N(1) THEN 32Ø\IF J>N(2) THEN 32Ø
31Ø  GOTO 23Ø
```

The allocation is achieved using the north-west corner method – for each depot and customer, the smaller of the depot's stock S(I,1) and the customer's order S(J,2) is allocated. When the depot's stock is smaller (line 260), the customer's order is reduced by this amount and the next depot considered with the same customer (line 270). The depot's stock is not set to zero because the depot will not be used again in this section of the program. Similar considerations apply to lines 280 and 290. A complication arises when the stock and the order are exactly equal (line 240). In this case a small allocation is given to the next customer in order to avoid a problem known as **degeneracy**. For the algorithm to work properly, the number of non-zero allocations made must be exactly equal to N(1)+N(2)−1 and this small allocation guarantees this. The subroutine called calculates the cost of the allocation C.

```
1100  C=C+X(I,J)*C(I,J)
1110  RETURN
```

The values at this stage are tabulated below.

CUSTOMER

DEPOT	1	2	3	4	5
1	7 / 4	8 / 2	0 / 1	0 / 2	0 / 0
2	0 / 3	2 / 5	4 / 4	0 / 3	0 / 0
3	0 / 5	0 / 5	4 / 2	13 / 1	15 / 0

Cost 91

The initial deliveries X(I,J) are shown on the top left of each square and the costs C(I,J) on the bottom right. This solution may not be the optimum and to check this a set of shadow costs is determined. The assumption made is that the transportation cost for all non-zero allocations can be expressed as the sum of a nominal depot cost and a nominal customer cost. These nominal or shadow costs will be stored in the lists S(I,1) and S(J,2). The first depot is given a shadow cost of 0, i.e. S(1,1)=0. From the non-zero allocations in the table, the shadow cost for customer 1 must therefore be S(1,2)=4, for customer 2, S(2,2)=2, for depot 2, S(2,1)=3 and so on. These shadow costs must be evaluated in the correct order – the program achieves this by finding a path from row 1 to the first column with a non-zero allocation, then to the next row with a non-zero allocation and so on.

```
320  FOR K=1 TO 2\FOR L=1 TO N(K)
330  P(L,K)=0\NEXT L\NEXT K
340  T(0)=0\T(1)=1\S(1,1)=0
350  K=2\Y=1\Z=1
360  GOSUB 1200
370  S(I(L),L)=C(I(1),I(2))-S(I(K),K)
380  Y=Y+1\IF Y < N(1)+N(2) THEN 360
```

The array P is defined as follows:
P(I,1) next column to be checked when moving along row I.
P(J,2) next row to be checked when moving down column J.
The **stack** T stores the rows and columns on the path – T(1)=1 indicates that row 1 is the first. Y stores the number of shadow costs determined so far, Z the number of rows and columns on the path, i.e. number of elements in the stack T and K is a flag – when K=1 a path is being made across a row and when K=2 down a column.

```
1200  K=3-K\I(K)=T(Z)\L=3-K
1210  P(I(K),K)=P(I(K),K)+1
1220  I(L)=P(I(K),K)
1230  IF I(L)>N(L) THEN 1270
1240  IF I(L)=T(Z-1) THEN 1210
1250  IF X(I(1),I(2))=0 THEN 1210
1260  Z=Z+1\T(Z)=I(L)\GOTO 1280
1270  Z=Z-1\GOTO 1200
1280  RETURN
1290  END
```

The contents of I(1) are the current row under consideration and those of I(2) the current column. The subroutine works as follows:

Line 1200 Since K=2 initially, this line sets K to 1, i.e. moving along a row. On the next call, K will be reset to 2 to move down a column and so on. I(K) is the row (column) which is currently on top of the stack. L is the flag for a column when K=1 and a row when K=2.
Line 1210 The next column (row) to check is determined.
Line 1220 This column (row) is now stored in I(L).
Line 1230 Check for end of row (column).
Line 1240 If the column (row) is already the penultimate one on the stack, try another column (row). This check prevents the path from retracing its steps.
Line 1250 If a move has been found to a zero allocation, try another column (row).
Line 1260 A path has been found to the next column (row). Place it on the stack and return.
Line 1270 End of row (column), i.e. the path cannot be continued. Remove the last row (column) from the stack and try to find a new path from the previous column (row).

The subroutine will find a path from row 1 to column 1 first giving a stack of T(1)=1, T(2)=1. It is then impossible to find a path down the first column, so line 1270 is executed to leave a stack of T(1)=1 – the value of Z indicates the number of relevant entries in the stack. The subroutine then finds a path from row 1 again to column 2, then to row 2, column 3, row 3, column 4 and then has to **unstack** this last column before finally reaching column 5. On each return from the subroutine when a new cell is reached, line 370 determines the shadow cost of the column (row). When the number of shadow costs found is N(1)+N(2) the program moves on to the next stage.

```
390 M=-1E6
400 FOR I=1 TO N(1)\FOR J=1 TO N(2)
410 Z=S(I,1)+S(J,2)-C(I,J)
420 IF Z<M THEN 440
430 M=Z\I1=I\J1=J
440 NEXT J\NEXT I
450 IF M<=0 THEN 660
```

The above code finds the sum of the two shadow costs minus the transportation cost for each cell in the table. The maximum value of these is found and stored in M, I1 and J1 storing the corresponding cell's row and column number. The data is given in the following table:

		4	2	1	0	−1	S(J,2)
S(I,1)	DEPOT	1	2	3	4	5	CUSTOMER
0	1	0	0	0	−2	−1	
3	2	4	0	0	0	2	
1	3	0	−2	0	0	0	

Thus M=4, I1=2, J1=1. When M is zero or negative, the optimum solution is the present allocation and the answer can be printed. Here the solution can be improved by allocating units of the product to the new cell. From the table on page 96, two units can be placed in cell (2,1), two units subtracted from cells (1,1) and (2,2) and two units added to cell (1,2). This operation ensures that the row and column totals remain the same. The cells to alter can be determined by finding the unique path from row I1 to column J1.

```
460 FOR K=1 TO 2\FOR L=1 TO N(K)
470 P(L,K)=0\NEXT L\NEXT K
480 T(1)=I1\K=2\Z=1
490 GOSUB 1200
500 IF K=2 THEN 490\IF T(Z)<>J1 THEN 490
```

The array P is zeroed and the path subroutine called again. The difference here is that row I1 is placed in the stack initially and the routine terminated when the last entry on the stack is column J1 (line 500). The first condition on this line is used to check whether the last entry is a row. K=2 indicates that the subroutine has just moved from a column to a new row. The stack at the end of this stage will be T(1)=2, T(2)=2, T(3)=1, T(4)=1. T(1) and T(3) are the rows, T(2) and T(4) the columns. The path is row 2→ column 2→ row 1→ column 1. The maximum allocation to cell (2,1) is now found.

```
510 M1=1E6\T(Z+1)=I1\Y=0
520 FOR K=1 TO Z-1 STEP 2
530 I=T(K)\J=T(K+1)
540 IF X(I,J)>M1 THEN 560
550 M1=X(I,J)
```

```
560 NEXT K
570 FOR K=1 TO Z-1 STEP 2
580 I=T(K)\J=T(K+1)
590 X(I,J)=X(I,J)-M1
600 IF X(I,J)>0 THEN 620\Y=Y+1
610 IF Y<=1 THEN 620\X(I,J)=0.001
620 I=T(K+2)
630 X(I,J)=X(I,J)+M1
640 NEXT K
650 C=C-M*M1\GOTO 320
```

The cells which are to have their allocations reduced are (2,2) and (1,1) and those to be increased are (1,2) and (2,1). The first K loop gives:

K	I	J	X(I,J)	M1
1	2	2	2	2
3	1	1	7	2

Here the cells (2,2) and (1,1) are checked and M1 set to the smaller of the allocations.

The second K loop then adjusts the allocations:

K	I	J	
1	2	2	X(2,2)=2−2=0
–	1	–	X(1,2)=8+2=10
3	1	1	X(1,1)=7−2=5
–	2	–	X(2,1)=0+2=2

Note: T(5) has been set to 2 on line 510.

Lines 600, 610 are added to avoid any further degeneracy. The procedure must only allocate a zero value to one cell and Y counts the number of such zero allocations. If Y ever reaches the value 2 or more, a small allocation is given to the corresponding cell on line 610. The cost of the new allocation is given by line 650. The shadow costs are then redetermined and the procedure repeated until the optimum solution is found. The last part of the program prints the optimum solution (see line 450) and updates the files.

```
660 N(F)=N(F)-1
670 FOR I=1 TO N(1)\FOR J=1 TO N(2)
680 X(I,J)=INT(X(I,J)+0.5)
690 IF X(I,J)=0 THEN 710
700 PRINT "SEND";X(I,J);"UNITS FROM SUPPLIER";F(I,1);
705 PRINT "TO CUSTOMER";F(J,2)
710 NEXT J\NEXT I
720 PRINT "COST";INT(C+0.5)
```

Line 660 removes any dummy depot or customer (see line 190). Line 680 rounds the allocations to the nearest integer in case degeneracy has occurred – an allocation of 0.001 is made purely for the correct working of the algorithm and represents a zero allocation in practical terms. For all non-zero allocations the results are printed but the number of the depot and customer *in the file* is printed on line 700.

```
730 FOR K=1 TO 2\RESTORE#K
740 OPEN #3, "TEMP"\SCRATCH #3
750 I(K)=1\I(3-K)=N(3-K)+1
760 IF END #K THEN 810\INPUT#K:Z
770 IF Z>0 THEN 790
780 PRINT#3,Z\GOTO 760
790 PRINT#3,X(I(1),I(2))
800 I(K)=I(K)+1\GOTO 760
810 CLOSE#K\CLOSE#3
820 OPEN #K, A$(K)\OPEN #3, "TEMP"
830 SCRATCH#K
840 IF END#3 THEN 870
850 INPUT #3, Z\PRINT #K,Z
860 GOTO 840
870 CLOSE#K\CLOSE#3
880 NEXT K
890 STOP
```

When K is 1, the depot file is reset to the beginning and a temporary file opened. This file has its contents deleted first because an empty file is needed here. I(1) on line 750 gives the number of the first depot in the program and I(2) the number of the dummy customer (if any). The contents of the depot file are then inspected. When a value of zero is read, this is printed into the temporary file (line 780) and the next number in the file is checked. In the latter case, the depot in question must have had zero stocks initially and has not been used in the program. For a non-zero value, line 790 prints the new stock, after the customers' orders have been removed, into the temporary file. In the case of a dummy customer, the new stock level is the amount sent to this imaginary customer by the depot (see I(2) above). When no dummy customer exists, X(I(1),I(2)) is zero anyway because I(2) represents the next column after the end of the table and is zeroed on line 100. The next depot in the program can then be processed (line 800). When the end of the depot file is reached, both files are closed on line 810. They are then re-opened to be used as an input (output) file instead of an output (input) file. The depot file is deleted on line 830 and the contents of the temporary file copied back into it. The customer's file is then updated in a similar fashion when K=2. Note that I(K) cannot be used as a loop variable in the program because it is a subscripted variable and Basic does not permit this.

This program will also solve the Classical Assignment Problem (see *Operations Research* in this series) if values of one for all stocks and orders are placed in the relevant files, though there are better algorithms that can be used in such a problem. This program does not allow the contents of the cost file to be altered due to increased overheads nor the addition or deletion of a depot or customer. By adding further sections of code to the program, the user can be offered various **menus**, i.e. a set of functions which the program will perform on command.

Chapter 13

Applications

LINEAR PROGRAMMING

A company produces three products A, B, C which yield a contribution of $5, $4, $6 respectively. The company wishes to implement the weekly production plan that maximizes profits. The following constraints must be satisfied:

PRODUCT	LABOR (HRS) PER UNIT	MATERIALS (LB) PER UNIT	SPECIAL COMPONENTS PER UNIT	PRODUCTION LIMIT
A	1	3	3	0
B	1	2	1	0
C	1	1	0	1
MAXIMUM AVAILABLE	100	200	90	40

The problem is a standard linear programming exercise, though it is not necessary to have prior knowledge of the Simplex method to understand the ideas behind the program. A demonstration of the method is given before each section. The first step is to set up the inequalities assuming that the weekly production of A, B, C is x_1, x_2, x_3 respectively.

Objective Function: maximize $5x_1+4x_2+6x_3$
Subject to $x_1+x_2+x_3 \leq 100$ (Labor)
$3x_1+2x_2+x_3 \leq 200$ (Materials)
$3x_1+x_2 \leq 90$ (Components)
$x_3 \leq 40$ (Production Limit)
and $x_1 \geq 0$, $x_2 \geq 0$, $x_3 \geq 0$.

The data can be input as follows:

```
10  DIM A(20,40)
20  PRINT "LP-MAXIMIZING BY SIMPLEX TABLEAU"
30  REM ZERO THE ARRAY
40  FOR I=1 TO 20\FOR J=1 TO 40
50  A(I,J)=0
60  NEXT J\NEXT I
70  REM INPUT THE DATA
80  PRINT "NUMBER OF CONSTRAINTS";\INPUT M
90  PRINT "NUMBER OF PRODUCTS";\INPUT N
100 L=M+N
110 FOR I=1 TO M
120 PRINT "CONSTRAINT NUMBER";I
130 FOR J=1 TO N
140 PRINT "NO. UNITS NEEDED PER UNIT OF PRODUCT";J;
150 INPUT A(I,J)
160 NEXT J
170 PRINT "MAXIMUM NO. UNITS AVAILABLE";
180 INPUT A(I,L+1)
```

```
190 NEXT I
200 FOR J=1 TO N
210 PRINT "CONTRIBUTION OF PRODUCT";J;
220 INPUT A(M+1,J)
230 NEXT J
```

For the example shown M=4, N=3, L=7. The equations stating that the variables are positive need not be entered and are guaranteed anyway by the method of solution. The array A(I,J) now contains the allocations:

A(1,1)=1,	A(1,2)=1,	A(1,3)=1,	A(1,8)=100	(inequality 1)
A(2,1)=3,	A(2,2)=2,	A(2,3)=1,	A(2,8)=200	(inequality 2)
A(3,1)=3,	A(3,2)=1,	A(3,3)=0,	A(3,8)= 90	(inequality 3)
A(4,1)=0,	A(4,2)=0,	A(4,3)=1,	A(4,8)= 40	(inequality 4)
A(5,1)=5,	A(5,2)=4,	A(5,3)=6,	A(5,8)= 0	(objective function)

Room has been left in the array to allow the addition of slack variables. These replace the inequalities by equalities:

$$\begin{aligned} x_1 + x_2 + x_3 + x_4 &= 100 \\ 3x_1 + 2x_2 + x_3 + x_5 &= 200 \\ 3x_1 + x_2 + x_6 &= 90 \\ x_3 + x_7 &= 40 \end{aligned}$$

The program will use the Simplex algorithm in the form of a table. The total number of variables is now L=M+N=7 because an additional variable is added to each of the constraints. The array has to be extended to include the allocations A(1,4)=1, A(2,5)=1, A(3,6)=1, A(4,7)=1 where A(I,J) is the coefficient of variable J in equation I. The remaining coefficients are already zero because of the zeroing of the array between lines 40 and 60.

```
240 REM ADD SLACK VARIABLES
250 FOR I=1 TO M
260 A(I,N+I)=1
270 NEXT I
```

The data is usually set out as the table:

Solution variable	x_1	x_2	x_3	x_4	x_5	x_6	x_7	Solution quantity	
x_4	1	1	1	1	0	0	0	100	First equation
x_5	3	2	1	0	1	0	0	200	Second equation
x_6	3	1	0	0	0	1	0	90	Third equation
x_7	0	0	1	0	0	0	1	40	Fourth equation
5	4	6	0	0	0	0	0		Objective function

The first and last columns represent the initial feasible solution to the problem, $x_4=100$, $x_5=200$, $x_6=90$, $x_7=40$, the remaining variables being zero. The last entry in the table represents an initial objective function value of 0. The solution variable column has also to be remembered in the program and will be stored in the array as A(1,0)=4,

A(2,0)=5, A(3,0)=6, A(4,0)=7, only the number of the variable being relevant.

```
28Ø REM ADD SOLUTION VARIABLE
29Ø FOR I=1 TO M
3ØØ A(I,Ø)=N+I
31Ø NEXT I
```

Now an attempt is made to improve the initial solution. The coefficients in the objective function are inspected and the largest one selected – here it is 6 under variable number 3. When the largest value is negative or zero, the Simplex algorithm ends and the solution printed.

```
32Ø REM FIND LARGEST COEFFICIENT
33Ø X=-1E99\C=Ø
34Ø FOR J=1 TO L
35Ø IF A(M+1,J)<X THEN 37Ø
36Ø X=A(M+1,J)\C=J
37Ø NEXT J
38Ø IF X<=Ø THEN 64Ø
```

X is the largest coefficient and C the corresponding variable number so that X=6, C=3 in the example. Line 38Ø transfers control to the printing of the answer when required. The third variable can now be increased to improve the value of the objective function. To determine its new value, the coefficient of x_3 in each of the equations is divided into the corresponding solution quantity and the minimum of these chosen. If the coefficient is negative or zero the equation is ignored, i.e.
Equation 1 gives 100/1=100
Equation 2 gives 200/1=200
Equation 3 gives 90/0 and is ignored
Equation 4 gives 40/1=40
The minimum value is 40 from Equation 4.

```
39Ø REM FIND NEW VALUE OF VARIABLE C
4ØØ X=1E99\R=Ø
41Ø FOR I=1 TO M
42Ø IF A(I,C)<=Ø THEN 46Ø
43Ø Y=A(I,L+1)/A(I,C)
44Ø IF Y>X THEN 46Ø
45Ø X=Y\R=I
46Ø NEXT I
```

This will find the minimum value X=40 and the corresponding equation number R=4. Problems can occur when R or X are zero but these complications will be ignored here. The third variable (C) can now be increased in value. The next step is to alter the fourth equation (R=4) in the table. The solution variable in this equation is changed to C=3 and the equation divided by the pivot element A(R,C)=A(4,3)=1 to give a fourth row of x_3 0 0 1 0 0 0 1 40.

The only change here is to the solution variable but the program has to perform the division in all possible cases.

```
470 REM CHANGE SOLUTION VARIABLE
480 A(R,0)=C
490 REM CHANGE REST OF ROW R
500 X=A(R,C)\FOR J=1 TO L+1
510 A(R,J)=A(R,J)/X
520 NEXT J
```

The next procedure is the most difficult. The remaining rows have to be changed so that the coefficient of variable C is zero in each of them. For example the last row is treated as follows:

ROW 5	5	4	6	0	0	0	0	0
6 x NEW ROW 4	0	0	6	0	0	0	6	240
SUBTRACT	5	4	0	0	0	0	−6	−240

A multiple of row R=4 is always subtracted from each row. Note that the new row has a zero coefficient for variable C=3. The multiple required in each case is simply the coefficient of this variable in the row under consideration. The solution variables must retain their original values.

```
530 REM CHANGE OTHER ROWS
540 FOR I=1  TO M+1
550 IF I=R THEN 610
560 X=A(I,C)
570 FOR J=1 TO L+1
580 A(I,J)=A(I,J)-X*A(R,J)
590 NEXT J
600 A(I,C)=0
610 NEXT I
```

Line 550 ensures that row R is not altered again. X is the required multiple for row I and the J loop performs the subtractions for each element. Line 600 has been added to make certain that A(I,C) is exactly zero in the event of the computer giving it a very small non-zero value (this may occasionally happen because of rounding errors). It is crucial for the remainder of the algorithm that this allocation is performed accurately. The table now becomes:

Solution variable	x_1	x_2	x_3	x_4	x_5	x_6	x_7	Solution quantity
x_4	1	1	0	1	0	0	−1	60
x_5	3	2	0	0	1	0	−1	160
x_6	3	1	0	0	0	1	0	90
x_3	0	0	1	0	0	0	1	40
−	5	4	0	0	0	0	−6	−240

The new solution can be read off from the solution variable and quantity columns as x_4=60, x_5=160, x_6=90, x_3=40, the remaining variables being zero. The value of the objective function is now 240 ignoring the negative sign of the last number in the array. Another iteration is now performed on the new array until all the coefficients in the objective function are negative or zero.

```
620 REM REPEAT ITERATION
630 GOTO 320
```

The final solution is reached via line 380 of the program. The answer is contained in the first and last columns of the array (omitting the last row). These columns may contain slack variables which are not printed.

```
640 REM PRINT SOLUTION
650 PRINT "OPTIMUM SOLUTION"
660 FOR J=1 TO N
670 X=0
680 FOR I=1 TO M
690 IF A(I,0)<>J THEN 710
700 X=A(I,L+1)\GOTO 720
710 NEXT I
720 PRINT "PRODUCE";X;"UNITS OF PRODUCT";J
730 NEXT J
740 X=-A(M+1,L+1)
750 PRINT "MAXIMUM PROFIT $";X
760 END
```

For each of the original variables J in the problem, a search is made among the solution variables A(I,0) for the value J. If a match is found, X is set to the solution quantity A(I,L+1) and the result printed. No match is found when the variable J is not present among the solution variables and then it has the value zero in the optimum solution (line 670). The final value of the objective function is the negative of the last entry in the array A(M+1,L+1).

```
OPTIMUM SOLUTION
PRODUCE 15 UNITS OF PRODUCT 1
PRODUCE 45 UNITS OF PRODUCT 2
PRODUCE 40 UNITS OF PRODUCT 3
MAXIMUM PROFIT $ 495
```

MULTI-SERVER QUEUES

The queuing problem of Chapter 7 is designed to simulate a single server situation. In this section multiple service points will be allowed and the relative merits of using various numbers of these will be considered. Customers may leave without being served if the queue length exceeds a number Q1 input by the user. The design of the queuing process in programming terms is different and more efficient than the first queuing example.

```
10 DIM S(10),T(10),W(10)
20 RANDOMIZE
30 PRINT "MAXIMUM NO. SERVICE PTS.";\INPUT N
40 PRINT "LENGTH OF SIMULATION(HOURS)";\INPUT T
50 PRINT "MEAN INTER-ARRIVAL TIME(SECS)";\INPUT A
60 PRINT "MEAN SERVICE TIME(SECS)";\INPUT S
70 PRINT "MAXIMUM QUEUE LENGTH";\INPUT Q1
```

```
80  PRINT "COST OF SERVICE PT. PER HOUR($)";\INPUT C1
90  PRINT "CONTRIBUTION PER CUSTOMER($)";\INPUT C2
1ØØ N=INT(ABS(N))
```

Line 1ØØ illustrates a method of ensuring that N is a positive whole number. With a sensible input of 5, N remains the same but when −3.7 is input, N=INT(ABS(−3.7))= INT(3.7)=3. The costs are added on lines 8Ø and 9Ø in order to determine the profit. For example a garage might wish to know if it would be profitable to install a new gas pump – in this case C1 would include part of the running costs, employees' salaries and installation costs, etc., and C2 would represent the average profit to the garage from each customer served.

```
11Ø FOR K=1 TO N
12Ø C=Ø\Q=Ø\E=Ø\L=Ø
13Ø FOR I=1 TO 1Ø
14Ø S(I)=Ø\T(I)=Ø\W(I)=Ø
15Ø NEXT I
```

The loop variable is the number of service points to be simulated and begins with one service point, then two and so on up to the maximum N. The simple variables zeroed are the clock time C in seconds, the length of the queue Q (not including customers being served), the number of arrivals E and the number of customers leaving without service L. For each service point I, S(I) is the number of customers served, T(I) the time, relative to the clock, of the end of service and W(I) the time that the service point is not in use.

```
16Ø Z=A\GOSUB 59Ø\T(Ø)=Z
17Ø W=T(Ø)
18Ø FOR I=1 TO K
19Ø IF T(I)<=Ø THEN 21Ø\IF T(I)>W THEN 21Ø
2ØØ W=T(I)
21Ø NEXT I
22Ø C=C+W\IF C>36ØØ*T THEN 37Ø
23Ø FOR I=Ø TO K\T(I)=T(I)−W\NEXT I
```

The subroutine called on line 16Ø takes the mean inter-arrival time stored in Z and returns an inter-arrival time at random now stored in Z. This is then placed in the variable T(Ø) ready for further analysis. The next section determines the time W to the next event, i.e. an arrival or the end of a service. For example:
T(Ø)=5, i.e. next arrival due in 5 seconds' time.
T(1)=3, i.e. service point 1 due to become free in 3 secs.
T(2)=−1. When the time is zero or negative the service point is already free.

The next event is the end of service here, the minimum of the times T(Ø), T(1), . . . , T(K) discounting any negative or zero values. W will then contain the value 3 at the end of the loop between lines 18Ø and 21Ø. The clock time is then advanced by W and the results are printed when this exceeds the length of the simulation in seconds. Line 23Ø alters the list T to give all times relative to the new clock time, e.g. T(Ø)=2, T(1)=Ø, T(2)=−4.

```
24Ø  IF T(Ø)>Ø THEN 27Ø
25Ø  E=E+1\Q=Q+1
26Ø  Z=A\GOSUB 59Ø\T(Ø)=Z
27Ø  FOR I=1 TO K
28Ø  IF T(I)>Ø THEN 34Ø\IF T(I)<Ø THEN 3ØØ
29Ø  S(I)=S(I)+1\GOTO 31Ø
3ØØ  W(I)=W(I)+W
31Ø  IF Q=Ø THEN 34Ø
32Ø  Z=S\GOSUB 59Ø\T(I)=Z
33Ø  Q=Q-1
34Ø  NEXT I
```

When T(Ø) is greater than zero the next arrival is not due yet. If it is zero, an arrival has occurred, E and Q are incremented and the next inter-arrival time sampled. The service points are then checked by the I loop:

T(I)>Ø End of service not due yet.
T(I)<Ø Service point has been free, update idle time, check queue.
T(I)=Ø End of service, i.e. service point now free, increment S(I), check queue.
Q=Ø Queue empty – no customers to be served.
Q>Ø Sample next service time, reduce queue length. The sampling of a new service time is performed by the same subroutine as that for the arrivals but Z is set to the mean service time S instead.

```
35Ø  IF Q<=Q1 THEN 17Ø
36Ø  L=L+Q-Q1\Q=Q1\GOTO 17Ø
```

The number of customers leaving the queue is now evaluated. When Q>Q1, the number leaving is Q−Q1 giving a new queue length of Q1. Line execution moves to line 17Ø to determine the next event.

```
37Ø  PRINT K;"SERVICE POINT(S)"
38Ø  FOR I=1 TO 18\PRINT "-";\NEXT I
39Ø  PRINT\PRINT E;"ARRIVALS"
4ØØ  Z=Ø
41Ø  FOR I=1 TO K\Z=Z+S(I)\NEXT I
42Ø  PRINT Z;"CUSTOMERS SERVED"
43Ø  PRINT L;"CUSTOMERS LEFT WITHOUT SERVICE"
44Ø  PRINT Q;"CUSTOMERS IN QUEUE"
45Ø  C=Ø
46Ø  FOR I=1 TO K
47Ø  IF T(I)<=Ø THEN 49Ø
48Ø  C=C+1
49Ø  NEXT I
5ØØ  PRINT C;"CUSTOMERS BEING SERVED"
51Ø  PRINT "TOTAL CONTRIBUTION $";Z*C2
52Ø  PRINT "TOTAL RUNNING COSTS $";T*C1*K
525  PRINT
53Ø  PRINT "SERVICE PT.   NO. SERVED   TIME IDLE(MINS)"
54Ø  FOR I=1 TO K
55Ø  PRINT TAB(6);I;TAB(19);S(I);TAB(31);W(I)/6Ø
56Ø  NEXT I
57Ø  NEXT K
58Ø  STOP
```

The total number of customers served is the sum of the entries in the list S(I). A customer is still being served at the end of the simulation if T(I) is greater than zero (line 48Ø). As a check to the program, E should equal the sum of Z, L, Q and C. The next value of K is then executed by returning to line 12Ø and zeroing all necessary variables before starting the new simulation.

```
59Ø  R=RND(Ø)
6ØØ  IF R=Ø THEN 59Ø
61Ø  X=Z*LOG(1/R)
62Ø  X=INT(X+Ø.5)
63Ø  IF X=Ø THEN 59Ø
64Ø  Z=X\RETURN
65Ø  END
```

The subroutine produces a random inter-arrival or service time from the negative exponential distribution – a theoretical distribution often used in queuing problems. The random number R must not be zero (line 6ØØ). X (line 61Ø) is the actual time and this is rounded to the nearest integer on line 62Ø. The program assumes that these times are non-zero and line 63Ø ensures that another time is sampled in this case. If the mean inter-arrival or service times are small the program should be changed to deal with intervals of a fraction of a second rather than an interval of one second.

Sample output when N=3, T=8, A=60, S=120, Q1=10, C1=15 and C2=1 are input.

```
2 SERVICE POINT(S)
------------------
483 ARRIVALS
439 CUSTOMERS SERVED
36 CUSTOMERS LEFT WITHOUT SERVICE
6 CUSTOMERS IN QUEUE
2 CUSTOMERS BEING SERVED
TOTAL CONTRIBUTION $ 439
TOTAL RUNNING COSTS $ 24Ø

SERVICE PT.     NO. SERVED      TIME IDLE (MINS)
     1             222              37.2833
     2             217              48.9333
```

The results for two service points only are shown here. The optimum number of service points to give a profit and maintain customer satisfaction can then be determined from the results of the simulation.

Chapter 14

Further Techniques

INDEX SORTING

The usual method of sorting data into some order can be a long and time-wasting exercise when other lists have also to be exchanged, e.g.

No.	NAME	AGE	YEARS OF SERVICE	DEPT.
1	SMITH J	18	1	DISPATCH
2	LANG J	42	12	SALES
3	CLARK F	31	3	SECRETARIAL
4	FEARN A	24	2	SALES

The lists have to be ordered alphabetically by name. To exchange the first two names needs the four lines:

```
100  D$=N$(1)\N$(1)=N$(2)\N$(2)=D$
110  D=A(1)\A(1)=A(2)\A(2)=D
120  D=Y(1)\Y(1)=Y(2)\Y(2)=D
130  D$=D$(1)\D$(1)=D$(2)\D$(2)=D$
```

The lists N$, A, Y and D$ store the names, ages, years of service and departments respectively.
Similarly other entries have to be exchanged in this way as part of a bubble-sort routine. The alternative is to use an **index list** X(I) which gives the number of the name in position I in the final order.

```
100  REM INDEX SORT
110  FOR I=1 TO N\X(I)=I\NEXT I
120  FOR I=1 TO N-1
130  FOR J=I+1 TO N
140  K=X(I)\L=X(J)
150  IF N$(K)<N$(L) THEN 170
160  X(I)=L\X(J)=K
170  NEXT J
180  NEXT I
```

N is 4 in the example and the index list is set as X(1)=1, X(2)=2, X(3)=3, X(4)=4 initially, e.g. X(2)=2 because the number of the name in position 2 of the initial order is name number 2. The loop begins by comparing the name in position I with the name in position J in the current index list. When the names are already in alphabetical order no change is made. Otherwise the relevant elements in the list are exchanged.

I	J	K	L	N$(K)	N$(L)	X(1)	X(2)	X(3)	X(4)
–	–	–	–	–	–	1	2	3	4
1	2	1	2	SMITH J	LANG J	2	1	3	4
1	3	2	3	LANG J	CLARK F	3	1	2	4
1	4	3	4	CLARK F	FEARN A	3	1	2	4
2	3	1	2	SMITH J	LANG J	3	2	1	4
2	4	2	4	LANG J	FEARN A	3	4	1	2
3	4	1	2	SMITH J	LANG J	3	4	2	1

The final index list is 3, 4, 2, 1 indicating that the third name is the first in alphabetical order, the fourth name the second and so on. The names can be printed by the following lines.

```
190 FOR I=1 TO N
200 K=X(I)
210 PRINT N$(K)
220 NEXT I
```

Dry run

I	K	N$(K)
1	X(1)=3	N$(3)="CLARK F"
2	X(2)=4	N$(4)="FEARN A"
3	X(3)=2	N$(2)="LANG J"
4	X(4)=1	N$(1)="SMITH J"

INVERSE OF A LIST

The index list X(I) in the last section gives the number of the name in position I. Sometimes a new list Y(J) is needed giving the position in the list of name J., e.g.

X(1)=3	Name 3 is first	Y(1)=4	Fourth is name 1
X(2)=4	Name 4 is second	Y(2)=3	Third is name 2
X(3)=2	Name 2 is third	Y(3)=1	First is name 3
X(4)=1	Name 1 is fourth	Y(4)=2	Second is name 4

The conversion can be achieved by:

```
230 FOR I=1 TO N
240 Y(X(I))=I
250 NEXT I
```

e.g.

I	X(I)	Y(X(I))=I
1	3	Y(3)=1
2	4	Y(4)=2
3	2	Y(2)=3
4	1	Y(1)=4

The expression Y(X(I)) is the element in position X(I) in the list Y. This bracketing of variables may also be used with arrays as in A(X(I), Y(J)): X(I) is the row number, Y(J) the column number and the whole expression the element in this row and column.

INDEX SELECTION

A list of people in the sales department can be extracted and stored from the index list X by the following:

```
260 P=0
270 FOR I=1 TO N
280 J=X(I)
290 IF D$(J)<>"SALES" THEN 320
300 P=P+1
310 X(P)=J
320 NEXT I
330 FOR I=1 TO P
340 J=X(I)
350 PRINT N$(J)
360 NEXT I
```

Dry run

I	J	P	Line 310
1	3	0	–
2	4	1	X(1)=4
3	2	2	X(2)=2
4	1	2	–

I	J	N$(J)
1	4	FEARN A
2	2	LANG J

P is used to count the number of people in sales and also acts as a pointer to the next position in the index list. This list now contains the entries X(1)=4, X(2)=2, X(3)=2, X(4)=1 because the last two were not altered by the selection procedure. The value of P must be taken into account here and indicates that only the first two entries are part of the selection. These entries are guaranteed to be in alphabetical order because they were already sorted. When the original index list X must be saved, the selected entries can be stored in another index list Z by replacing lines 31Ø and 34Ø by Z(P)=J and J=Z(I) respectively. Alternatively selection can be made from the original list of names and then sorted afterwards if necessary.

CRITICAL PATH ANALYSIS

The next program will determine the critical activities of a network using the framework of an Activity on Nodes diagram. The conventional method of drawing networks using events as the nodes is not particularly suited to analysis by a computer. The list A$(I) stores the names of the activities, a maximum of 50 activities with up to 8 characters in each name being allowed for. The relevant times are stored in the array T(I,J) defined as folows:
T(I,Ø) the duration of activity I.
T(I,1) the earliest start time (EST) of activity I.
T(I,2) the latest start time (LST) of activity I.
The earliest finish time (EFT) for activity I is then T(I,1)+T(I,Ø) and the latest finish time (LFT) T(I,2)+T(I,Ø).

The storage of preceding activities may be accomplished in at least three different ways. For example:

Activity	No.	Preceding Activities	Preceding Activity Nos.	Duration T(IØ)
A	1	E, D	5, 4	3
B	2	D	4	4
C	3	–	–	2
D	4	C	3	5
E	5	–	–	1
F	6	B, A	2, 1	4

The activities need not be input in the correct order and the start and finish activities are added by the program. The array P(I,J) could be used to store this table as P(I,J)=1 if activity J immediately precedes activity I and P(I,J)=0 otherwise. Then P(1,1)=0, P(1,2)=0, P(1,3)=0, P(1,4)=1, P(1,5)=1, P(1,6)=0, etc., giving a total of 36 locations in this example. An alternative method defines Q(I,J) as the Jth preceding activity of activity I so that Q(1,1)=5, Q(1,2)=4, Q(2,1)=4, Q(2,2)=0, the zero indicating no second activity. This needs 6×2=12 locations in memory here although this will vary in other examples depending upon the maximum number of preceding activities that an activity has. Finally two lists A(I) and B(I) may be employed as follows: A(1)=5, B(1)=1, i.e. 5 precedes 1; A(2)=4, B(2)=1, i.e. 4 precedes 1 and so on for every preceding activity. This requires 12 locations in memory and is probably the most efficient method of storage. The number of locations in memory required for each of the three examples is not the only factor which determines the best one to use. The program itself uses computer memory, and storage for program variables could be considerably reduced if the program is too long. The preference on this basis would be to use the array P, then the array Q and finally the two lists. Another factor to be taken into consideration is the time that the program needs to find a preceding activity from the arrays or lists. The array P would have to be searched to find entries of one and the lists searched to find a particular activity number. The array Q gives the preceding activities immediately but it is still difficult to determine whether an activity is a preceding activity or not. Searches can be speeded up by storing more data in arrays so that relevant information is defined explicitly though the problem of storage becomes a crucial factor in a network with 500 activities or more. In this program the first method will be employed to reduce the amount of coding under the assumption that there are not too many activities. An example of storage in the form of the array Q is given in the next section of this chapter.

```
10  PRINT "CRITICAL PATH ANALYSIS"
20  DIM A$(51,8),T(51,2),P(51,51),X(51),F(51)
30  REM ZERO THE ARRAYS
40  FOR I=0 TO 51
50  FOR J=0 TO 2\T(I,J)=0\NEXT J
60  FOR J=0 TO 51\P(I,J)=0\NEXT J
70  NEXT I
80  REM INPUT THE DATA
90  PRINT"UP TO 50 ACTIVITIES MAY BE INPUT IN ANY ORDER"
100 PRINT"DO NOT INCLUDE START AND FINISH ACTIVITIES"
110 PRINT "NUMBER OF ACTIVITIES";
120 INPUT N
130 FOR I=1 TO N
140 PRINT "ACTIVITY NAME";
150 INPUT A$(I);
160 PRINT "DURATION";
170 INPUT T(I,0)
180 NEXT I
190 FOR I=1 TO N
200 PRINT "ACTIVITY";A$(I)
210 PRINT "NUMBER OF PRECEDING ACTIVITIES";
220 INPUT M
230 IF M=0 THEN 340
240 PRINT "INPUT THE PRECEDING ACTIVITIES"
250 FOR P=1 TO M
260 INPUT Z$
270 FOR K=1 TO N
280 IF Z$<>A$(K) THEN 300
290 J=K\GOTO 320
300 NEXT K
310 PRINT "NAME NOT KNOWN-INPUT AGAIN"\GOTO 260
320 P(I,J)=1
330 NEXT P
340 FOR K=1 TO 57\PRINT"-";\NEXT K
350 PRINT
360 NEXT I
```

The dimension statement has to allow for a maximum of 52 activities because the start activity is coded as zero and the finish activity as N+1 where N is the number of activities input. The K loop between lines 270 and 300 searches for the name of the preceding activity Z$ in the list of names A$(I) in order to determine its number J. P(I,J) can then be set to one indicating that activity J is a predecessor of activity I. When the name Z$ is not found, the user is asked to change the last input on line 310. The other K loop on line 340 prints a line of dashes before moving on to the next activity. The array P will now contain the non-zero entries P(1,5), P(1,4), P(2,4), P(4,3), P(6,1) and P(6,2) in the example given.

The activities have now to be sorted in order that the analysis of the times is performed correctly. The ordering must satisfy the rule that all predecessors of an activity must appear earlier in the order than the activity itself. In the example, A has predecessor E which appears later in the list showing that E must be analyzed before A. A bubble-sort is difficult to apply because it is not always obvious whether two activities should be interchanged or not. The sorting algorithm used here starts by finding activities with no predecessors and places them in the index list X before searching for other activities which can then be placed. The flag list F is defined as F(I)=1 if activity I has been positioned correctly and F(I)=0 otherwise.

```
370  REM SORT THE ACTIVITIES
380  FOR I=1 TO N\X(I)=I\F(I)=0\NEXT I
390  K=1
400  FOR P=K TO N
410  I=X(P)
420  FOR J=1 TO N
430  IF P(I,J)=0 THEN 450
440  IF F(J)=0 THEN 470
450  NEXT J
460  P1=P\GOTO 490
470  NEXT P
480  PRINT "ERROR IN NETWORK"\GOTO 1080
490  X(P1)=X(K)\X(K)=I
500  F(I)=1\K=K+1
510  IF K<>N THEN 400
```

X(P) gives the activity in position P and this list and the flags are initialized on line 380. The first pass through the P loop begins with P=1, and line 410 finds the activity I currently in position P in the index list. The J loop determines the predecessors J of this activity indicated by P(I,J)=1. If a predecessor is found whose flag is still zero, the next activity in the index list is considered (line 440). Line 460 is only executed when an activity I has no predecessors or when all its predecessors have their flags set to one indicating that they have all been positioned correctly. In the latter cases, the current position of activity I is stored in P1. Line 490 then swaps the activities in positions P1 and K of the index list, sets the flag of activity I to 1 and increases K by 1. For the example, the first activity to be placed correctly will be activity 3 in position 3, i.e. P1=3, I=3. Lines 490, 500 then give X(P1)=X(K)=X(1)=1, i.e. X(3)=1, X(K)=I, i.e. X(1)=3, F(3)=1, K=2. The index list is now X(1)=3, X(2)=2, X(3)=1, X(4)=4, X(5)=5, X(6)=6 indicating that the first position is now fixed for activity 3 but the remainder have yet to be processed. It does not matter that activity 1 has been relegated to third place at the moment. The next P loop will now begin with the activity in position 2 of the new list and will be able to place activity number 4 because its sole predecessor, activity 3, has just been placed (F(3)=1). This process is then repeated until the first N−1 activities have been placed (i.e. K=N on line 510), implying that the remaining activity must also be in the correct position. The index list takes the following values during the process:

P	X(P)	X(P)	X(P)	X(P)	X(P)
1	3	3	3	3	3
2	2	4	4	4	4
3	1	1	2	2	2
4	4	2	1	5	5
5	5	5	5	1	1
6	6	6	6	6	6

The final order is then 3, 4, 2, 5, 1, 6, i.e. C, D, B, E, A, F which is adequate to cope with the calculation of the start and finish times of each activity. If execution ever reaches line 480, the P loop must have been performed without any activities being placed, signifying an error in the network.

```
520 REM ADD START ACTIVITY AS NO. ZERO
530 A$(0)="START"\X(0)=0
540 FOR I=1 TO N\FOR J=1 TO N
550 IF P(I,J)=1 THEN 580
560 NEXT J
570 P(I,0)=1
580 NEXT I
```

The duration of the start activity is already zero. It is given position zero in the index list. The double loop checks each activity I to see if it has a predecessor when P(I,J)=1 for some value of J. If not, the start activity must be its predecessor and line 570 updates the array P to give P(3,0)=1, P(5,0)=1 in the example.

```
590 REM ADD FINISH ACTIVITY AS NO. N+1
600 A$(N+1)="FINISH"\X(N+1)=N+1
610 FOR J=1 TO N\FOR I=1 TO N
620 IF P(I,J)=1 THEN 650
630 NEXT I
640 P(N+1,J)=1
650 NEXT J
```

The finish activity is given position N+1 in the index list. The double loop this time checks each activity J to see if it is ever a predecessor when P(I,J)=1 for some value of I. If not, it must be a predecessor of the finish activity. The array then has the additional entry P(7,6)=1 because activity F is the only one which is never a predecessor in the original table.

```
660 REM EST AND EFT
670 T(0,1)=0
680 FOR P=1 TO N+1
690 S=0\I=X(P)
700 FOR Q=0 TO P-1
710 J=X(Q)
720 IF P(I,J)=0 THEN 760
730 F=T(J,1)+T(J,0)
740 IF F<S THEN 760
750 S=F
760 NEXT Q
770 T(I,1)=S
780 NEXT P
```

The EST of the start activity is set to zero on line 670. The forward pass through the positions P of the index list finds the activity I in this position and sets its EST, S, to zero initially. The EST is the maximum of the earliest finishing times of its preceding activities. Because of indexing, these preceding activities must be in a position earlier than that for activity I. The Q loop enumerates the relevant positions to give any possible predecessors J. When P(I,J) is zero, J is not a predecessor and the next position is checked. Otherwise F, the EFT of activity J, is evaluated and placed in S if it is the largest value so far. When all predecessors have been found, the EST for activity I, T(I,1), must have this latter value.

```
790 REM LST AND LFT
800 T(N+1,2)=T(N+1,1)
810 FOR P=N TO 0 STEP -1
820 F=1E6\J=X(P)
830 FOR Q=P+1 TO N+1
840 I=X(Q)
850 IF P(I,J)=0 THEN 890
860 S=T(I,2)
870 IF S>F THEN 890
880 F=S
890 NEXT Q
900 T(J,2)=F-T(J,0)
910 NEXT P
```

This section is very similar to the previous one. The LST of the finish activity must be equal to its EST calculated by the previous section. A backward pass is required because the LFT of an activity J is the minimum of the latest start times of the activities for which J is a predecessor. F represents this LFT and is set large on line 820. The positions to be checked this time are all later than that for activity J. The variable S contains the LST of activity I when J is one of its predecessors. On line 900 the LST of activity J is its LFT minus its duration. Because of the index list the EST and LST of an activity are never evaluated before the relevant values for neighboring activities are known.

```
920  REM PRINT THE RESULTS
930  PRINT "ACTIVITY";TAB(11);"EST";TAB(16);"EFT";
940  PRINT TAB(21);"LST";TAB(26);"LFT";TAB(31);"SLACK"
950  FOR I=0 TO N+1
960  PRINT A$(I);
970  F(1)=T(I,1)\F(2)=F(1)+T(I,0)
980  F(3)=T(I,2)\F(4)=F(3)+T(I,0)
990  F(5)=F(3)-F(1)
1000 FOR J=1 TO 5
1010 PRINT TAB(5*J+6);F(J);
1020 NEXT J
1030 IF F(5)<>0 THEN 1050
1040 PRINT TAB(38);"CRITICAL";
1050 PRINT
1060 NEXT I
1070 STOP
```

The EST, EFT, LST, LFT and the slack time are stored in the first five places of the flag list before printing.

```
1080 REM ERROR IN NETWORK
1090 IF K=1 THEN 1170
1100 PRINT "THESE ACTIVITIES HAVE BEEN ORDERED"
1110 FOR P=1 TO K-1
1120 I=X(P)
1130 PRINT A$(I)
1140 NEXT P
1150 PRINT "NO OTHERS CAN BE ADDED"
1160 STOP
1170 PRINT "EVERY ACTIVITY HAS A PREDECESSOR"
1180 END
```

This part of the program is executed when the sorting of the activities has failed. If K is still 1, the ordering has never begun because an activity without a predecessor has not been found. Otherwise the error has occurred when trying to find the activity to place in position K and the results of the order so far are printed. The commonest network errors occur when a set of activities form a loop, the sorting routine being unable to position any of them.

Sample Output

ACTIVITY	EST	EFT	LST	LFT	SLACK	
START	0	0	0	0	0	CRITICAL
A	7	10	8	11	1	
B	7	11	7	11	0	CRITICAL
C	0	2	0	2	0	CRITICAL
D	2	7	2	7	0	CRITICAL
E	0	1	7	8	7	
F	11	15	11	15	0	CRITICAL
FINISH	15	15	15	15	0	CRITICAL

ASSIGNMENT PROBLEMS

A company wishes to set up a new production process which requires the completion of five independent tasks.

Five outside contractors have submitted time estimates for each task as follows:

		Job 1	2	3	4	5	
	1	12	9	6	3	5	
	2	11	7	5	5	4	
Contractor	3	7	6	6	4	3	Time in months
	4	8	9	5	8	4	
	5	8	8	7	5	3	

The company wishes to assign each contractor one task in such a way that the production process is on line as soon as possible. The contractors perform their tasks simultaneously so that the time to reach production is the longest of the times needed to finish a job. When there is a choice between assignments with the same completion time, the one with the smallest sum of times is to be selected. For example the following two possible assignments have the same completion time of seven months:

Contractor	1	2	3	4	5	Contractor	1	2	3	4	5
Job	4	2	1	5	3	Job	5	2	1	3	4
Time	3	7	7	4	7	Time	5	7	7	5	5

The first solution is preferred because it has a smaller sum of times than the second. This type of problem is called a bottleneck assignment in which the maximum of the times needed to complete a job has to be made as small as possible. The algorithm required to solve the problem is far simpler than that needed for the classical assignment problem when the first criterion is to ensure that the sum of the times is a minimum. Prior knowledge of bottleneck assignments is not necessary to understand the ideas behind the program.

```
10 PRINT "BOTTLENECK ASSIGNMENT"
20 DIM T(20,20),S(20,20),R(20),P(20),F(20),A(20)
30 REM ZERO THE TWO ARRAYS
40 FOR I=1 TO 20\FOR J=1 TO 20
50 T(I,J)=0\S(I,J)=0
60 NEXT J\NEXT I
70 REM INPUT THE DATA
80 PRINT "NUMBER OF JOBS";\INPUT M
90 PRINT "NUMBER OF CONTRACTORS";\INPUT N
100 IF M=N THEN 160\IF M<N THEN 120
110 PRINT "ASSIGNMENT IMPOSSIBLE"\GOTO 80
120 FOR J=M+1 TO N
130 PRINT J;
140 NEXT J
150 PRINT "ARE THE DUMMY JOBS"
160 FOR I=1 TO N\FOR J=1 TO M
170 PRINT "CONTRACTOR"; I; "JOB"; J; "TIME";
180 INPUT T(I,J)
190 NEXT J\NEXT I
```

When the number of jobs M is greater than the number of contractors N, an assignment cannot be made because the program assumes that each contractor can do at most one job. When M is less than N, the problem can be solved by adding a number of dummy jobs which take zero time to complete. This ensures that the number of jobs is equal to the number of contractors. In the example N is 5 and the time for contractor I and job J is stored as T(I,J). An algorithm to find the best assignment proceeds as follows:

1. Find the smallest time(s) in the table.
2. Try to find an assignment using these times only.
3. If step 2 fails, find the next smallest time.
4. Goto step 2.

For example, smallest time is 3 for contractor 1 job 4, contractor 3 job 5 and contractor 5 job 5. Store the results:

S(I,Ø)=number of jobs included so far for contractor I, i.e. S(1,Ø)=1, S(3,Ø)=1, S(5,Ø)=1, the remainder being zero.

S(I,1)=the first job included for contractor I, i.e. S(1,1)=4, S(3,1)=5, S(5,1)=5, the remainder being zero.

F(J) is a flag which is set to 1 when job J has been included, i.e. F(4)=1, F(5)=1 (the flag list is zeroed before the algorithm begins). An allocation is not possible because only three contractors and two jobs have been stored. Next smallest time is 4 for contractor 2 job 5, contractor 3 job 4 and contractor 4 job 5. These are now added to the array S and the flag list F to give:

Contractor I	No. jobs S(I,Ø)	First job S(I,1)	Second S(I,2)
1	1	4	0
2	1	5	0
3	2	5	4
4	1	5	0
5	1	5	0

F(1)=0, F(2)=0, F(3)=0, F(4)=1, F(5)=1

An allocation is still impossible because some jobs are not represented yet.

```
200 FOR J=1 TO N\F(J)=Ø\NEXT J
210 M=-1
220 M1=1E12
230 FOR I=1 TO N\FOR J=1 TO N
240 IF T(I,J)<=M THEN 270
250 IF T(I,J)>M1 THEN 270
260 M1=T(I,J)
270 NEXT J\NEXT I
280 M=M1
290 FOR I=1 TO N\FOR J=1 TO N
300 IF T(I,J)<>M THEN 340
310 P=S(I,Ø)\P=P+1
320 S(I,P)=J\S(I,Ø)=P
330 F(J)=1
```

```
340 NEXT J\NEXT I
350 REM CHECK IF ASSIGNMENT POSSIBLE
360 FOR I=1 TO N
370 IF S(I,Ø)=Ø THEN 22Ø
380 NEXT I
390 FOR J=1 TO N
400 IF F(J)=Ø THEN 22Ø
410 NEXT J
```

M is now used to store the last smallest time (the number of jobs and contractors are both N). To find the next smallest time M1, the array T is examined and all entries less than or equal to M are excluded from the search on line 24Ø. Otherwise the problem is a standard exercise in finding the minimum of the remaining entries. M is set to −1 initially to ensure that all entries in the array are checked on the first pass through this part of the program. When the new minimum M1 is found, M can then be set to the same value in readiness for any further execution of this section. The next loop from lines 29Ø to 34Ø finds the entries with this time and stores the new entries. P is the number of jobs included for contractor I so far and this is incremented by 1 before placing job J in the next position in the list of his jobs. S(I,Ø) has then to be given this new value of P and the flag set for job J. Note that a flag may be set more than once during the loop and a conditional statement to avoid this would not give different results. An assignment is certainly impossible when a contractor has not been included (line 37Ø) or a job has not been included (line 4ØØ). In these cases the process has to be repeated by returning to line 22Ø.

The next step is to see if an assignment is now possible when the program reaches line 42Ø. This line will be executed when the array S(I,J) contains the following values:

Contractor I	S(I,Ø)	S(I,1)	S(I,2)	S(I,3)	S(I,4)	S(I,5)
1	3	4	5	3	–	–
2	4	5	3	4	2	–
3	5	5	4	2	3	1
4	2	5	3	–	–	–
5	3	5	4	3	–	–

The order of the jobs for each contractor is not important for the correct execution of the algorithm. The value of M is 7 at this stage and all the flags are set to 1. It is easy to see some possible assignments from this table, e.g. the contractors can do jobs 5, 2, 1, 3, 4 respectively but a method of reaching this answer has to be carefully broken down into a set of well-defined instructions. A technique known as **backtrack** will be used here:

I is the number of the contractor being considered.
I=1 Contractor 1 can do job 4, the first on his list – store the assignment
R(1)=4, i.e. job 4 allocated to contractor 1
P(1)=1, i.e. the first job on the list has been used
F(4)=1, i.e. job 4 assigned (all flags are reset to zero first)

I=2 Contractor 2 can do job 5, the first on his list.
R(2)=5, i.e. job 5 allocated to contractor 2
P(2)=1, i.e. the first job on the list has been used
F(5)=1, i.e. job 5 now assigned
I=3 The first job, 5, and the second job, 4, have been used (flags are 1).
Contractor 3 can do job 2, the third on his list
R(3)=2, P(3)=3, F(2)=1
I=4 This gives R(4)=3, P(4)=2, F(3)=1
I=5 Contractor 5 cannot be given a job because all jobs on his list have already been assigned.
The algorithm then moves back to the fourth contractor (i.e. backtracks) as follows:
Reset P(5) to 1, i.e. when contractor 5 is considered again, the search will begin from his first job.
Set P(4) to 3 (one more than before), i.e. examine the third job (if any) on the fourth contractor's list.
Reset F(3) to zero, i.e. job 3 cannot now be allocated (at this stage) to contractor 4.
I=4 and P(4)=3. There is no third job on the list. Backtrack to contractor 3.
Reset P(4) to 1, i.e. search from first job when I=4 again.
Set P(3) to 4 (one more than before), i.e. examine fourth job for I=3.
Reset F(2) to zero, i.e. job 2 cannot now be allocated to the third contractor.

The process is then repeated for I=3 and job 3, the fourth on the list, will be assigned to contractor 3 before moving onto contractor 4 again. Many further backtracks will be needed before a solution is reached. At each stage an assignment is attempted without upsetting, if possible, the previous assignments already made. The list P contains pointers to determine the position of the next job to be examined when a backtrack needs to be made. This algorithm guarantees that all possible allocations will be found and will not retrace its steps to check assignments which have already been determined, thus entering an infinite loop.

```
420 T=1E12\F=0
430 FOR I=1 TO N
440 R(I)=0\F(I)=0\P(I)=1
450 NEXT I
460 I=1
470 J=S(I,P(I))
480 IF F(J)=1 THEN 520
490 R(I)=J\F(J)=1
500 IF I=N THEN 570
510 I=I+1\GOTO 470
520 P(I)=P(I)+1
530 IF P(I)<=S(I,0) THEN 470
540 IF I=1 THEN 660
550 P(I)=1\I=I-1
560 F(R(I))=0\GOTO 520
```

Line 420 T is used to store the smallest sum of times for the assignments found.
F is another flag which is set to 1 when a complete assignment is found.
Line 440 The variables are initialized – all pointers set to position 1.
Line 460 Considering the first contractor.
Line 470 J is the job for contractor I which is in position P(I).
Line 480 If job assigned, try next job on list (line 520).
Line 490 Job J can be allocated, set flag.
Line 500 When I=N the assignment must be complete.
Line 510 Otherwise move onto next contractor and find job J again.
Line 520 Try the next job on the list by increasing the position by 1.
Line 530 If there is a next job, find it and test its flag.
Line 540 No next job. When I=1, backtrack is impossible and no more assignments exist.
Lines 550 and 560
Reset pointer, backtrack to previous contractor, free the job R(I) for further use and go to 520 to check the next job on the list for this new contractor.

```
570  REM ASSIGNMENT FOUND
580  F=1\T1=0
590  FOR K=1 TO N
600  T1=T1+T(K,R(K))
610  NEXT K
620  IF T1>T THEN 650
630  T=T1
640  FOR K=1 TO N\A(K)=R(K)\NEXT K
650  F(R(N))=0\GOTO 550
660  IF F<>0 THEN 690
670  FOR J=1 TO N\F(J)=1\NEXT J
680  GOTO 220
690  REM SOLUTION
700  FOR I=1 TO N
710  PRINT "CONTRACTOR";I;"JOB";A(I);"TIME";T(I,A(I))
720  NEXT I
730  PRINT "TIME NEEDED";M
740  PRINT "SUM OF TIMES";T
750  END
```

When an assignment is found (I=N), the flag F is set to 1, and the sum of the times T1 for this is calculated by the loop from lines 590 to 610. K is used as the loop variable because there may be other possible assignments to be found and I is already being used in the backtrack section. K is the contractor and R(K) the job he has been assigned. T stores the smallest sum and, when required, the best assignment so far is stored in the list A(K) on line 640. A return is then made to search for any other assignments by backtracking again. The next line is executed when no more assignments exist (see line 540). If the flag F is still zero, the backtrack procedure has found no assignments at all. This would happen in the following example:

Contractor	Jobs
1	4,5
2	5,3,4
3	5,4,2,3,1
4	5,3
5	5,4,3

Contractor 3 is the only one who can do jobs 1 and 2.

Control then returns to line 220 to find the next smallest time in the array T before trying the whole backtrack procedure again. The flags F(J) have to be set to 1 in this case to indicate that all jobs are represented. The list F(J) is being used for two different purposes in the program. The remaining possibility on line 660 is that flag F is 1 when no more assignments exist but some have already been found. The solution can then be printed, the job for each contractor I being stored in the list A(I). There are four possible assignments in the example giving a minimum completion time of seven months:

Contractor	1	2	3	4	5	TIMES	SUM
1 Job	4	2	1	5	3	3,7,7,4,7	28
2 Job	4	2	1	3	5	3,7,7,5,3	25
3 Job	5	2	1	3	4	5,7,7,5,5	29
4 Job	3	2	1	5	4	6,7,7,4,5	29

The assignment chosen by the program will be the second with the lowest value in the last column. Another method of choosing the best assignment is to find those with the smallest completion time and then select the one with the smallest cost attached to it. The bottlenecks in the solution are jobs 1 and 2 because they both have an estimated time of 7 months and must be completed on time to avoid delay in the start of the production process. On the other hand, job 4 with a completion time of only 3 months can take as much as 4 months longer without delaying the whole project. The illustration used is an easy one because of the small number of jobs – the program becomes useful when the number of jobs is of the order of 10 or more.

Further Reading

Booth, T.L., *Digital Networks and Computer Systems* (Wiley, second edition, 1978)
Hall, P.A.V., *Computational Structures* (MacDonald & Jane's American Elsevier Inc., 1975)
Kemeny, J.G. and Kurtz, T.E., *Basic Programming* (Wiley, third edition. 1980)
Lindley, D., *Making Decisions* (Wiley, 1975)
Lucey, T., *Quantitative Techniques* (D.P. Publications, 1979)
Maynard, J., *Computer Programming* (W.H. Allen, 1972)
Tremblay, J., Bunt, R.B. and Opseth, L.M., *Structured Pascal* (McGraw-Hill, 1980)

Glossary

Algorithm A finite set of well-defined steps to solve a problem.
Alphabetic data Information in the form of the letters of the alphabet.
Alphanumeric data Information whose characters can be numbers, letters or punctuation symbols.
Alphanumeric file A file storing alphanumeric information.
Applications program A program which performs specific user oriented tasks, e.g. all programs in this book.
Argument A constant, variable or string variable from which a function determines its value.
Array A collection of data with the structure of a list, table, etc.
ASCII A specific code assigning a number to each alphanumeric character.
Assembler A systems program which translates an assembly language into machine code.
Assembly language A low level language corresponding closely to a machine language.
Assignment statement A program statement which defines the contents of a variable.
Backtrack programming A method of systematically enumerating a set of possible allocations.
Binary search A search procedure which halves the number of elements to be scanned at each stage.
Branch command A statement allowing the programmer to pass control to another part of a program.
Bubble sort A sorting procedure which "bubbles" elements up a list until a smaller value is reached.
Central processor That part of a computer which performs all arithmetic and logic operations and sends signals to peripheral devices.
Coding The translation of a problem into a programming language.
Compiler A systems program which translates a high level language into a machine language.
Compile time diagnostic An indication of any program errors encountered during the compilation of the program.
Computer A machine which processes data to produce information in another form.
Computer run The output given by the computer during the running of a program.
Concatenation The process of combining strings.
Conditional transfer Passing control to some other part of a program when a certain condition is satisfied.
Connector symbol A special symbol which indicates the path through a flow chart.
Control statement A statement which allows the order of line execution to be varied.
Control variable A variable whose value determines line execution.

Counter A variable used to count the number of elements in a list, array, etc.
Cumulative summation A method of addition where each number is successively added to the current total.
Data Information in the form of numbers, words or a combination of these.
Data file A list of information which can be stored on peripheral devices.
Debugging The process of locating bugs (errors) in a program.
Decision box A branching operation which determines the alternative paths to be followed in a flow chart.
Declaration The instructions to the computer to allocate space in memory for variables.
Directory file A file containing information relevant to the access of the main files in a program.
Double subscripted variables A set of variables used to store matrices and tables.
Dry run A run performed by hand to check the logic of a program.
Dump variable A variable used to store temporarily the contents of another variable in order to avoid loss of data.
Error code A mnemonic printed by the computer to signify a particular type of program error.
Exponentiation The operation of raising a number to some power.
False data An artificial value indicating the end of a data list or file.
FIFO A mnemonic for the operation of a queue – first in, first out.
Flag variable A variable whose value indicates the status of a condition.
Flow chart A network showing the logic of a program in a manner which is easy to follow.
Hard copy Output on paper from a line-printer or teletype, etc.
Hardware The mechanical and electrical components of a computer including any peripheral devices.
Heuristic A general principle that often leads to the correct solution of a problem (often used in games such as chess, etc.).
Hierarchy The order in which operations (both arithmetic and logical) are performed.
High level language A programming language which uses statements closely related to English, e.g. BASIC.
Index list An auxiliary list storing pointers to another list for the purpose of sorting and selection.
Inductive formula A formula defining the next value in a series in terms of previous values.
Infinite loop A set of lines in a program with no possible exit.
Information retrieval The process of extracting data from a file or data list.
Input device Any device allowing data to be fed into the computer.
Input file A file from which to read data.
Interpreter A systems program which converts a high level language into a machine language but whose operation varies from that of a compiler in that each statement is

initially translated and then executed in sequence.
Iterative process A method of performing a series of steps to produce successively better approximations to the solution of an equation or problem.
Jump instruction A command to pass control to another part of a program.
Key A variable giving the range of a set of data.
LIFO A mnemonic for the operation of a stack – last in, first out.
Linear search A search procedure which scans from the beginning to the end (or vice versa) of a list of items.
Link A pointer to the next element to be processed.
List A set of items in one dimension.
Location A part of memory where data or instructions may be stored.
Loop A set of lines in a program which are successively executed a given number of times.
Loop variable A variable controlling the operation of a loop.
Low level language A language which uses mnemonics closely corresponding to machine instructions or binary code.
Mainframe A computer with a large storage potential and the ability to process data at a very high rate.
Mapping function A construction giving the address of each element in an array.
Matrix A two dimensional array.
Menu One of the functions which a program may offer.
Merging The combination of two lists or arrays into a single structure.
Microcomputer Computers using microchips to reduce the size of the electronic components.
Minicomputer A smaller version of a mainframe computer.
Monte Carlo method Any procedure using random numbers, e.g. simulation.
Nested loop A loop placed inside another loop.
Nested subroutine A subroutine which is part of another subroutine.
Null string The empty string consisting of a single blank space.
Numeric data Data in the form of numbers, including the decimal point and the E format.
Numeric file A file containing numeric data.
Output device Any device which prints the output of a program.
Output file A file used to write data into.
Parameter A variable assigned a specific data value in order to avoid explicit mention of the value in the remainder of a program.
Peripheral device Any device external to the computer's memory and central processor.
Pointer A variable indicating the next element to be scanned.
Push down (up) list Another term for a stack (queue).
Queue A list in which the first element accessed is the first on the list.

Recursive formula A formula which defines the next value in terms of previous values.
Relation operator Any of the symbols =, >, <, >=, <=, <> used to compare numbers and strings (also applied to sets etc.).
Rogue value An artificial value signifying the end of a list of data.
Rounding errors Errors compounded due to the fact that computers can only store a certain number of digits for each number.
Run time diagnostic An indication of program errors encountered during the running of the program.
Semantic error An error of meaning (a logic error).
Serial file A file whose contents must be accessed by reading from the beginning of the file.
Simple variable A variable such as A, A1 defining a single location in memory.
Software The programs which give instructions to the computer, i.e. application programs and systems programs.
Stack A list in which the first element accessed is the last one on the list.
String A set of alphanumeric characters.
String constant A simple string variable such as A$ with no subscripts.
String function Any function whose argument or value is a string.
String variable A variable which stores a string.
Subroutine A section of code which can be called from any part of the main program when needed.
Subscript A number giving the row or column of an element in an array.
Subscripted variable A numeric or string variable which stores a particular element of an array.
Summation variable A variable which stores the sum of a set of numbers.
Switch variable A variable whose value determines line execution.
Syntax error An error in the use of language construction.
Systems program A program which controls the performance of the computer.
Top down design A method of coding which lists the functions of the program and uses subroutines to perform the analysis and input/output routines.
Transfer command Another term for control statement.
Truncation errors Errors caused by rounding down numbers or by omitting terms from a series.
Unconditional transfer A transfer to some other line of a program irrespective of any condition being satisfied.
Unstack To remove an element from the top of a stack.